SYMMETRIES
AND
ELEMENTARY PARTICLES

DOCUMENTS ON MODERN PHYSICS

A. ABRAGAM – *L'Effet Mössbauer*

H. BACRY – *Lectures on Group Theory*

T. A. BRODY – *Symbol Manipulation Techniques for Physics*

K. G. BUDDEN – *Lectures on Magnetoionic Theory*

J. W. CHAMBERLAIN – *Motion of Charged Particles in the Earth's Magnetic Field*

S. CHAPMAN – *Solar Plasma and Geomagnetism and Aurora*

H. CHIU – *Neutrino Astrophysics*

A. H. COTTRELL – *Theory of Crystal Dislocations*

J. DANON – *Lectures on the Mössbauer Effect*

BRYCE S. DEWITT – *Dynamical Theory of Groups and Fields*

R. H. DICKE – *The Theoretical Significance of Experimental Relativity*

M. GOURDIN – *Laws of Symmetry Theorem of T. C. P.*

D. HESTENES – *Space-Time Algebras*

JOHN G. KIRKWOOD – *Selected Topics in Statistical Mechanics*

JOHN G. KIRKWOOD – *Macromolecules*

JOHN G. KIRKWOOD – *Theory of Liquids*

JOHN G. KIRKWOOD – *Theory of Solutions*

JOHN G. KIRKWOOD – *Proteins*

JOHN G. KIRKWOOD – *Quantum Statistics and Cooperative Phenomena*

JOHN G. KIRKWOOD – *Shock and Detonation Waves*

JOHN G. KIRKWOOD – *Dielectrics–Intermolecular Forces–Optical Rotation*

R. LATTES – *Methods of Resolutions of Some Boundary Problems in Mathematical Physics*

FRANCIS E. LOW – *Symmetries and Elementary Particles*

PAUL H. E. MEIJER – *Quantum Statistical Mechanics*

M. MOSHINSKY – *Group Theory and the Many-body Problem*

A. B. PIPPARD – *The Dynamics of Conduction Electrons*

L. SCHWARTZ – *Applications of Distributions to the Theory of Elementary Particles in Quantum Mechanics*

Additional Volumes in preparation.

SYMMETRIES AND ELEMENTARY PARTICLES

FRANCIS E. LOW

Notes by Felix A. E. Pirani

GORDON AND BREACH, SCIENCE PUBLISHERS

New York • London • Paris

150 Fifth Avenue, New York, N. Y. 10011

Library of Congress Catalog Card Number 67–31587

Editorial Office for Great Britain:
Gordon and Breach, Science Publishers Ltd.
8 Bloomsbury Way
London, W. C. 1

Editorial Office for France:
Gordon & Breach
7–9 rue Emile Dubois
Paris 14^{e}

Distributed in France by:
Dunod Editeur
92 rue Bonaparte
Paris 6^{e}

Distributed in Canada by:
The Ryerson Press
299 Queen Street West
Toronto 2B, Ontario

Printed in the United States of America

F. E. LOW

Massachusetts Institute of Technology

0. PLAN OF THE LECTURES

These lectures are extremely elementary. They cover the symmetry principles which hold for the strong interactions.[1] Under this heading I shall discuss also the spectrum of the particles, and apply the symmetry principles to give an overall phenomenological picture of the strong interactions. Apart from some brief remarks today, I shall not discuss the symmetries associated with the Lorentz group or with gauge invariance. Those symmetries which I shall discuss in detail are space inversion (P), charge conjugation (C), time reversal (T), PCT, isotopic spin ($\vec{I}$), G-parity, and SU_3 (fairly completely, but without group theory).

1. SYMMETRY PRINCIPLES AND QUANTUM NUMBERS

Table 1 gives a summary of the symmetry principles and the associated quantum numbers, and some comments on

TABLE 1

Symmetry principle	Quantum numbers (constants of the motion)	Accuracy of conservation* (see text for further comments on each symmetry principle)
A. Continuous space-time symmetries		
1. Translation invariance	P_μ	Exact (for an isolated system)
2. Rotational invariance	$\vec{J}$	Exact
3. Lorentz invariance	$\vec{N}$ (boost)	Exact
B. Gauge symmetries		
4. General covariance	The equivalence principle $m_{\text{graviton}} = 0$ equations of motion of sources of the field.	Perhaps exact
5. Gauge invariance of electromagnetic field	Q, $\dot{Q} = 0$ (charge) $m_\gamma = 0$	Highly exact
6. Gauge invariance of a neutral vector field coupled to baryon number[2]	N(baryon number)	Highly exact
7. Gauge invariance of another strongly coupled field[3]	Y(hypercharge or strageness) S	Violated by weak† interactions
8. γ_5 invariance of the neutrino	$m_\nu = o$	
9. Gauge invariance of another vector field	lepton number	

*In this table the word "exact" is always to be understood qualified by "so far as present knowledge goes."

†The separation of interactions into strong, electromagnetic, weak and gravitational is a useful phenomenological device. We should keep in mind that it may be no more than that, since nature, unlike our theories, never turns off one or another of these interactions.

TABLE 1 (Cont'd.)

Symmetry principle	Quantum numbers	Accuracy of conservation
C. Discrete space-time symmetries		
10. Time reversal (T)	none (because the operator T is not unitary, and $[S, T] \neq 0$).	Uncertain Violated if CPT holds[4]
11. Space inversion (indistinginstability of an observation and of its mirror image).	P	Violated by weak interactions[5]
12. Charge conjugation (indistinguishability of a particle and its antiparticle).	C	Violated by weak interactions. Other levels of violation have been suggested.
13. CP($\sim$ T, if CPT is conserved).	CP	Violated in $K_2^0 \rightarrow 2\pi$ decay[4]
14. CPT (requires that particle and antiparticle have same mass even if C is violated).	CPT	If C strongly violated, then there is strong evidence for CPT from $K\bar{K}$ mass difference: $m_{k_1} - m_{k_2} \sim 10^{-4}$ ev, while $m_k \sim 5 \times 10^8$ ev, so that CPT holds to $\sim$ 1 in 10^{12}
D. Internal symmetries		
15. Charge independence	$\vec{I}^2$, I_3	Believed conserved by strong interactions; known violated by electromagnetic interactions.
16. G - parity invariance	$G = Ce^{-i\pi I_2}$	Conserved whenever $\vec{I}^2$ and C are both conserved.
17. Hypercharge independence (or "unitary symmetry," associated with SU_3)	8 conserved currents, namely I_1, I_2, I_3, Y and four others.	Strongly violated

the extent to which they are conserved. The symmetry principles are divided into four parts: continuous space time, gauge, discrete space time, and internal symmetries. I shall discuss the first two groups briefly and the last two in detail.

1.1. TRANSLATION INVARIANCE

A symmetry principle can usually be formulated as a statement about the impossibility of knowing something. Thus translation invariance states that all points in space-time are equivalent—there is no way of knowing where you are. Any inhomogenity observed so far has always been accounted for by identifying a field which produces it. Thus translational invariance applies to an isolated system.

Of course, from the point of view of cosmology, one takes into account the universe as a whole, and the idea of an isolated system has not an exact meaning. Here there is a difference in points of view. The elementary particle physicist would like to consider equations of motion decoupled from boundary conditions, and regard space-time as asymptotically flat and empty. The cosmologist would say that this must not be done—that there is no such thing as an isolated system—and that the actual universe must be brought into the boundary conditions, even that the laws of physics must include a specification of the universe, everything being fixed by mutual interactions—a sort of bootstrap picture of physics

1.2. ROTATIONAL INVARIANCE

Cosmological comments could also be made about this case. In addition, I should like to point out that actual experiments are always carried out in external fields— for example, the magnetic, electrostatic and gravitational fields of the earth. Experiments are not carried out to such an accuracy that the effects of these fields will show up. Even so, it is generally believed that translational and rotational invariance hold with a much greater accuracy than the limit which can be set by present terrestrial experiments.

1.3. LORENTZ INVARIANCE

The constants of the motion which arise in this case are not useful because they are explicitly time-dependent.

1.4. GENERAL COVARIANCE

I have suggested some consequences of general covariance, analogous to constants of the motion, which arise when general covariance is given physical content by requiring that the metric tensor be a dynamical variable and the equations for it generally covariant.

General covariance may hold exactly, but the experimental evidence for general relativity offered by the three classical tests is very slight: The redshift will come out of any linear theory which embodies the equivalence principle. The bending of light is not a consequence of the equivalence principle, but it is a linear, first-order effect. The precession of Mercury's perihelion is a second-order effect, but it is only one number.

Thus if we consider general relativity in successive orders of the gravitational potential, certainly to first order it is true, and the transformation involved in general covariance is a gauge-like transformation of the gravitational potential, but in the second order only one number has been predicted correctly.

We may ask what the theory says about the mass of the graviton. Presumably, if one knew how to translate general covariance into a self-mass calculation, then vacuum polarization graphs like

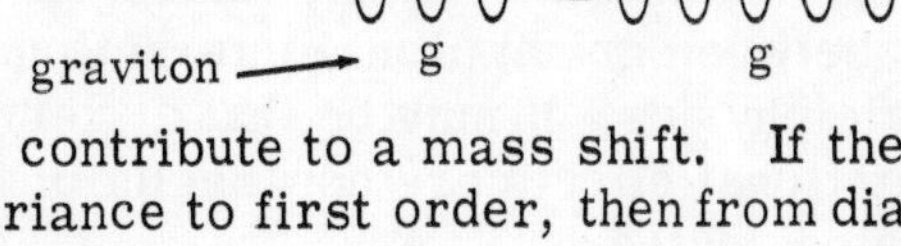

would not contribute to a mass shift. If there were no general covariance to first order, then from diagrams like this we could estimate the mass of the graviton, m_g, to be given by

$$m_g^2 \sim G^1 M^2$$

where G^1 is a suitable demensionless gravitational coupling constant and M is whatever mass comes into the diagram, say nucleon or electron mass. In natural units, we have

$$\frac{Gm_e^2}{e^2} \sim 10^{-42}, \text{ so presumably } G^1 \sim 10^{-42}.$$

Thus for first order violations of covariance $m_g \sim 10^{-21} m_e$. This gives a range for the gravitational force

$$r \sim 10^{11} \text{ cm}$$

which is much too small. The known range of the gravitational force is almost intergalactic, say 10^6 light years $\sim 10^{24}$ cm.

But if you go to second order, the graviton mass is given by

$$m_g^2 \sim G'^2 M^2$$

which gives for second-order violation

$$m_g \sim 10^{-42} m_e$$

and now

$$r \sim 10^{32} \text{ cm}$$

which is much greater than the radius of the universe.

We have to conclude from this that the observations require the vanishing of the graviton mass only in the first order, and that general covariance is almost not required by observation beyond the linear theory.

Another thing which emerges from this argument is that the gravitational constant is so small, compared even to the β-decay coupling constant, that so far almost no one has seen a connection between gravitational interactions and elementary particle physics. It may be that at very small distances gravitational effects are non-negligible, but at present they will be neglected.

1.5. GAUGE INVARIANCE OF THE ELECTROMAGNETIC FIELD

Besides the conservation of charge, shown in Table 1, there is of course for more complicated processes a whole sequence of relationships generated by gauge invariance, called 'generalized Ward identities,' each of which leads to a set of conserved quantities. Charge conservation is the most interesting.

Besides this, the photon mass must vanish (at least in a perturbation series in $\alpha = 1/137$)

$$m_\gamma = 0.$$

This is known very accurately. From observations of cosmical magnetic fields it appears that the Compton wavelength λ of the photon has a high lower limit:

$$\lambda \gtrapprox \frac{1}{10} \text{ light year} \sim 10^{18}\text{cm}$$

Experiments to measure the departure from neutrality of hydrogen and helium atoms[6] show that the proton and electron charges are the same and the neutron charge zero to 1 part in 10^{19}. The charge of the neutrino is comparably small. Thus charge is accurately conserved in β decays.

Random events in which charge is not conserved have never been observed; however some observational selection may be suspected in this connection!

1.6. GAUGE INVARIANCE COUPLED TO BARYON NUMBER

This is the first of a number of cases in which one knows the quantum number but not the symmetry principle. Yang and Lee[7] suggested that conservation of baryon number be accounted for by a field, analogous to the electromagnetic field, and coupled to baryon number N, whose gauge invariance would lead to $\dot{N} = 0$. At that time, it was believed, on the basis of perturbation theory, that gauge invariance could not be achieved except for a field of zero mass (the converse—that a zero mass field is impossible to achieve without some gauge invariance—is almost certainly true). Now a zero mass field would have infinite range. Since this field would be coupled to nucleon number and not to mass it would appear to violate the equivalence principle, and so a very low upper limit on the strength of the coupling would be imposed by the Eötvös-Dicke verification of the equivalence principle.[7]

However, it is now believed[8] that if the coupling is strong enough then gauge invariance does not necessarily imply zero mass. The ω (or ϕ?) meson is a candidate for this strongly coupled particle.

In this connection I should mention also experiments to measure the lifetime of the proton,[9] which show that at least with respect to leptonic decay modes it has a lifetime very much greater than the age of the universe. The persistence of matter indicates that the proton must be very stable.

1.7. GAUGE INVARIANCE COUPLED TO HYPERCHARGE

Sakurai[3] suggested that, again by analogy with electromagnetism, there should be a field strongly coupled to hypercharge. At the time, this idea was rejected for the same reasons as in the case of baryon number, but Sakurai persisted, and as we have seen, he was right—gauge invariance is possible for a strongly coupled field of non-zero mass (perhaps the ϕ (or ω?)).

As shown in the table, it is at present believed that hypercharge conservation is not violated in strong or electromagnetic interactions.

1.8. γ_5 INVARIANCE OF THE NEUTRINO

This is discussed by Prof. Cabibbo in his lectures. However, I should like to comment that at present the experimental evidence for γ_5 invariance comes from β-decay, while the mass of the ν is not known to be small enough to put any experimental limits on the γ_5 invariance: the dimensionless β-decay coupling constant is

$$G \sim 10^{-5}/M_N^2$$

(where M_N is the nucleon mass) so that the self mass of the neutrion would at best be

$$\sim 10^{-5} \text{ x (some mass)}$$

say

$$10^{-5}\ M_N \approx 10^{-2}\ \text{Mev}$$

which is larger than the known m_ν. However, when you carry out the calculation you find that the mass diagram vanishes because of helicity-invariance unless $m_e \neq 0$. It follows that all the mass effects must be multiplied by a small number like m_e/M_N, and then nothing is established experimentally. Therefore γ_5 invariance must be inferred from β-decay, not from the mass spectrum, at least until the experimental mass of ν_μ is brought down as low as that of ν_e, which might make other arguments possible.

1.9. GAUGE INVARIANCE COUPLED TO LEPTON CONSERVATION

There could be two more vector fields, one coupled to e and ν_e, one to μ and ν_μ. I am not competent to discuss the accuracy of this conservation law, so I shall leave the weak interactions there.

This completes our brief discussion of continuous space-time and gauge symmetries. The more detailed discussion of the remaining symmetries and conservation laws occupies the next two sections.

2. DISCRETE SPACE-TIME SYMMETRIES

In this section we study the discrete space-time symmetries P, C and T in some detail.

2.0. NOTATION

We work mainly in the Heisenberg picture. States in Hilbert space are denoted by capital Greek letters Φ, Ψ, ..., field operators by lower case Greek letters ϕ, ψ and capital Roman letters A, V, J, ... The notation for Dirac matrices agrees with that of Professor Cabibbo. With units chosen so that

$$\hbar = c = 1,$$

the free Dirac equation has the form

$$-\frac{1}{i}\frac{\partial\psi}{\partial t}(\vec{x}, t) = (\vec{\alpha}\cdot\frac{\vec{\nabla}}{i} + \beta m)\psi(\vec{x}, t). \tag{2.1}$$

A confusion of notation may arise from the fact that the spinor field ψ has four (or, in the non-relativistic case, two) components. To avoid this confusion let us use the following conventions: the components of ψ are labelled with a Greek index α; Greek indices range over 1, 2, 3, 4, and repeated indices are summed. In general we may think of the components of ψ as written in a column matrix

$$\psi = \begin{bmatrix} \psi_1 \\ \psi_2 \\ \psi_3 \\ \psi_4 \end{bmatrix},$$

but we also may use $\tilde{\psi}$ to denote the row matrix with the <u>same</u> components: $\tilde{\psi} = [\psi_1 \psi_2 \psi_3 \psi_4]$. In general, we denote a transposed matrix with ~. When ψ's and other matrices are written adjacently, matrix multiplication is always understood, and the indices suppressed. For example, the β in equation (2.1) denotes a 4×4 numerical matrix whose components $\beta_{\gamma\delta}$ will be specified shortly (equation (2.6)); in the Hilbert space β is merely a multiple of the unit operator, and to be very exact we should write $\beta\, 1$ instead of β. Accordingly $(\beta\psi)_\gamma$ is short for the set of four quantities

$$\sum_{\delta=1}^{4} \beta_{\gamma\delta}\,\psi_\delta,$$

which we abbreviate $\beta_{\gamma\delta}\,\psi_\delta$. On the other hand, we can also give a meaning to $\psi\beta$, namely the set of four quantities $\psi_\delta \beta_{\delta\gamma}$ (the order of the indices being determined by our convention that in such a product matrix multiplication is understood). Therefore, the equation $\beta\psi = \psi\tilde{\beta}$ is always true, whatever the components of β. It happens that in our case $\beta = \tilde{\beta}$, but this is accidental. By staying in the particular representation in which $\beta = \tilde{\beta}$, we gain in definiteness, although we may lose generality.

On the other hand, the adjoint of the operator ψ will be denoted $\psi^\dagger$. If (Ψ, Φ) denotes the inner product in Hilbert space, then $\psi^\dagger$ is defined by

$$(\Psi, \psi^\dagger \Phi) = (\psi\Psi, \Phi) \tag{2.2}$$

for all Ψ, Φ. But again $\psi^\dagger$ may denote the column matrix with entries

$$\begin{bmatrix} \psi_1^\dagger \\ \psi_2^\dagger \\ \psi_3^\dagger \\ \psi_4^\dagger \end{bmatrix}$$

or the row matrix with the same entries. In order to be able to construct Lorentz-invariant currents we shall make use of

$$\overline{\psi} = \psi^\dagger \beta = \tilde{\beta} \psi^\dagger = \beta\psi^\dagger, \tag{2.3}$$

which is the set of four quantities $\beta_{\gamma\delta} \psi_\delta^\dagger$.

The 4×4 Dirac matrices $\vec{\alpha}$ and β are direct products of 2×2 Pauli matrices $\vec{\rho}$ and $\vec{\sigma}$ defined by

$$\left.\begin{aligned} \vec{\alpha} &= \rho_1 \vec{\sigma} \\ \beta &= \rho_3 = \rho_3 1 \end{aligned}\right\} \tag{2.4}$$

where

$$\left.\begin{aligned} \sigma_x &= \begin{pmatrix} 0 & 1 \\ 1 & 0 \end{pmatrix} = \rho_1 \\ \sigma_y &= \begin{pmatrix} 0 & -i \\ i & 0 \end{pmatrix} = \rho_2 \\ \sigma_z &= \begin{pmatrix} 1 & 0 \\ 0 & -1 \end{pmatrix} = \rho_3 . \end{aligned}\right\} \tag{2.5}$$

The σ's are the spin-space matrices, the ρ's the charge-space matrices of the Dirac equation. Of course, the ρ's commute with the σ's. Even more explicitly, we may write

$$\vec{\alpha} = \begin{pmatrix} 0 & \vec{\sigma} \\ \vec{\sigma} & 0 \end{pmatrix} , \quad \beta = \begin{pmatrix} 1 & 0 \\ 0 & -1 \end{pmatrix} \tag{2.6}$$

(where 1 denotes the 2 × 2 unit matrix). We shall always use this representation of the Dirac matrices, in which the ρ's and σ's, and therefore also the α's and β's, are all self-adjoint. It will be called the "standard representation." Of course it is possible to carry out all calculations in a way which is independent of the representation, but it is frequently useful for practical purposes to be able to work things out in a particular representation, and we shall continually do this.

In momentum space, equation (2.1) becomes

$$(\vec{\alpha} \cdot \vec{p} + \beta m - E)\psi(p) = 0 \tag{2.7}$$

and in terms of covariant γ matrices this may be written

$$(i\gamma \cdot p + m)\ \psi = 0 \tag{2.8}$$

where

$$p = (\vec{p},\ iE) \tag{2.9}$$

and

$$\begin{aligned} \vec{\gamma} &= -i\beta\vec{\alpha} = \rho_2\vec{\sigma} \\ \gamma_4 &= \beta = \rho_3 \end{aligned} \tag{2.10}$$

are all self-adjoint matrices. Following Cabibbo we shall write

$$x_4 = ict$$

so that the 4th component of a real 4-vector is pure imaginary, and other related reality conditions must hold. The space-time scalar product is consequently

$$A \cdot B = \vec{A} \cdot \vec{B} + A_4B_4. \tag{2.11}$$

2.1. PARITY

In this section we shall calculate the transformation properties of various operators under space reflection, and see to what extent the assumption that reflection leaves field equations invariant is supported by experiment. In the following two sections we shall carry out a similar program for time reversal and charge conjugation. We work along parallel lines in these three sections, but distinctive features arise in each case, especially when we discuss experiments.

Let us consider now a Dirac field interacting with some other fields:

$$-\frac{1}{i}\frac{\partial\psi}{\partial t}(\vec{x},t) = \left[\vec{\alpha}\cdot\left(\frac{\vec{\nabla}}{i} - e\vec{A}\right) + eV + \beta(m + g_s\,\phi_s) + \rho_2\, g_p\,\phi_p\right]\psi(\vec{x},\ t) \tag{2.12}$$

where $(\vec{A},\ iV) = A_\mu$ is some (as yet unspecified) 4-vector potential, and the subscripts s an p denote 'scalar' and 'pseudoscalar' respectively.*

We introduce an operator P_{op} in the Hilbert space of the field operators, which we assume to be time-independent, and which reflects the space coordinates in ψ.

The invariance of equations (2.12) under this operation can be achieved by the assumption that P_{op} transforms ψ by a linear transformation (with constant coefficients) of its components. This will be shown immediately below. That such an assumption is necessary, as well as sufficient, may be shown to follow from considerations about the transformation properties of the Dirac matrices and the assumption that ψ transforms under a representation of the Lorentz group.

Accordingly, we assume that

$$P_{op}\,\psi_\alpha(\vec{x},t)\,P_{op}^{-1} = \eta_p\,P_{\alpha\beta}\psi_\beta(-\vec{x},\ t) \tag{2.13}$$

*The introduction of these various field operators ψ (four component), $\vec{A}$, V, ϕ_s, ϕ_p (and of course generalizations are possible) is not to be taken to imply that field theory must be used; this is just a convenient way of finding out the transformation properties of the corresponding particle states.

where $P_{\alpha\beta}$ is a set of numbers to be determined (a constant unitary matrix) and η_p is an undetermined phase factor. Suppressing the indices α, β in (2.13) gives the alternative form of this equation

$$P_{op}\,\psi(\vec{x},t)\,P_{op}^{-1} = \eta_p\, P\psi(-\vec{x},t) \qquad (2.13')$$

Operating on (2.12) with P_{op} (on the left) and P_{op}^{-1} (on the right), and multiplying on the left by the constant matrix P^{-1}, we find

$$-\frac{1}{i}\frac{\partial}{\partial t}\psi(-\vec{x},t) = P^{-1}\Big[\vec{\alpha}\cdot\Big(\frac{\vec{\nabla}}{i} - e\vec{A}'(\vec{x},t)\Big) + eV'(\vec{x},t)$$
$$+ \beta(m + g_s\phi_s'(\vec{x},t))$$
$$+ \rho_2 g_p \phi_p'(\vec{x},t)\Big]\,P\psi(-\vec{x},t) \qquad (2.14)$$

where

$\vec{A}'(\vec{x},t) = P_{op}\vec{A}(\vec{x},t)P_{op}^{-1}$ and V', ϕ_s' and ϕ_p' are similarly defined.

Changing $\vec{x}$ to $-\vec{x}$ throughout this equation, we find

$$-\frac{1}{i}\frac{\partial}{\partial t}\psi(\vec{x},t) = P^{-1}\Big[\vec{\alpha}\cdot\Big(-\frac{\vec{\nabla}}{i} - e\vec{A}'(-\vec{x},t)\Big) + eV'(-\vec{x},t)$$
$$+ \beta(m + g_s\phi_s'(-\vec{x},t))$$
$$+ \rho_2 g_p \phi_p'(-\vec{x},t)\Big]\,P\psi(\vec{x},t)$$

and if parity is to be conserved, this must be identical with the original equation (2.12). The comparison shows immediately that

$$P^{-1}\,\vec{\alpha}\,P = -\vec{\alpha}$$
$$P^{-1}\,\beta\,P = \beta; \qquad (2.15)$$

These equations have the solution, unique up to a phase,

$$P = \beta = \rho_3,$$

and substitution into (2.14) now gives

$$\vec{A}'(-\vec{x}, t) = -\vec{A}(\vec{x}, t)$$
$$V'(-\vec{x}, t) = V(\vec{x}, t)$$
$$\phi_s'(-\vec{x}, t) = \phi_s(\vec{x}, t)$$
$$\phi_p'(-\vec{x}, t) = -\phi_p(\vec{x}, t)$$

From (2.13) and (2.15) we can now deduce the transformation under reflections of the various currents which can be constructed from ψ and $\overline{\psi} = \psi^\dagger\beta$: let

$$\gamma_5 = \gamma_1\gamma_2\gamma_3\gamma_4 = -\rho_1$$

and notice that

$$P_{op}\,\psi(\vec{x}, t)\,P_{op}^{-1} = \eta_p\,\beta\,\psi(-\vec{x}, t)$$

implies

$$P_{op}\,\overline{\psi}(\vec{x}, t)\,P_{op}^{-1} = \eta_p^*\,\overline{\psi}(-\vec{x}, t)\,\beta.$$

Then we may calculate P_{op} (current) P_{op}^{-1}: the results are given in Table 2.

This completes our survey of parity transformation properties. We now consider the experimental validity of parity conservation:

In nuclear physics, one looks for a (p, γ) interaction involving two states of opposite parity, close together so that they mix easily, but with the electric transition inhibited in favour of the magnetic transition. There is some controversy, but it is stated that parity violation, of the order of magnitude expected from β-decay but certainly no larger, is observed. (By observing the circular polarization of the photon).

Now parity violation is observed in (weak) strangeness changing or leptons-producing interactions, for example in

$$N + \overline{N} \rightarrow \text{two leptons},$$

but one does not observe weak strangeness conserving interactions between baryons themselves.
Thus there are two possibilities:

TABLE 2

Current	Transformed current under P_{op}	Effect of transformation
Scalar $\bar{\psi} 1 \psi$	$\rightarrow \bar{\psi} \beta 1 \beta \psi = \bar{\psi} 1 \psi$	no change
Vector $\bar{\psi} \gamma_\mu \psi$	$\rightarrow \bar{\psi} \beta \gamma_\mu \beta \psi = \begin{pmatrix} -\bar{\psi} \vec{\gamma} \psi \\ \bar{\psi} \gamma_4 \psi \end{pmatrix}$	space components change sign time component unchanged
Tensor $\bar{\psi} \gamma_\mu \gamma_\nu \psi$	$\rightarrow \bar{\psi} \beta \gamma_\mu \gamma_\nu \beta \psi = \begin{pmatrix} \bar{\psi} \gamma_r \gamma_s \psi & -\bar{\psi} \gamma_r \gamma_4 \psi \\ -\bar{\psi} \gamma_4 \gamma_r \psi & \bar{\psi} \gamma_4 \gamma_4 \psi \end{pmatrix}$	space-space and time-time components unchanged others change sign
Axial vector $i \bar{\psi} \gamma_5 \gamma_\mu \psi$	$\rightarrow i \bar{\psi} \beta \gamma_5 \gamma_\mu \beta \psi = \begin{pmatrix} i \bar{\psi} \gamma_5 \vec{\gamma} \psi \\ -i \bar{\psi} \gamma_5 \gamma_4 \psi \end{pmatrix}$	space components unchanged time component changes sign
Pseudo-scalar $i \bar{\psi} \gamma_5 \psi$	$\rightarrow i \bar{\psi} \beta \gamma_5 \beta \psi = -i \bar{\psi} \gamma_5 \psi$	changes sign

1) Parity violating non-strangeness-changing weak interactions between baryons are of the same order as between baryons and leptons, with amplitudes proportional to $G/M_N^2 \sim 10^{-5}/M^2$. This corresponds to the current-current hypothesis, or,
2) Weak N. S. C. interactions between baryons are generated by weak interactions of the types already observed, and are therefore more likely to be of order

$$G^2 \sim 10^{-10}.$$

The experiment quoted above tends to favour the former possibility.

In 'elementary particle physics,' that is to say, using an arbitrary but possible definition, for systems with baryon number ≤ 1, it is harder to find tests of comparable accuracy. Probably the most accurate is the failure to observe the decay

$$\eta \rightarrow 2\pi$$

If parity were completely violated in strong interactions, the time for this decay would be comparable to the ρ lifetime, say $\sim 10^{-23}$sec. Presumably $\eta \rightarrow 2\pi$ occurs with only a few percent of the probability for $\eta \rightarrow 2\gamma$; of course, the time for this latter decay is unknown, but it may be comparable to that for $\pi \rightarrow 2\gamma$, say $\sim 10^{-16}$ sec. Therefore, a conservative estimate is that $\eta \rightarrow 2\pi$ fails by 1 in 10^6, or 1 in 10^3 in the amplitude. Finally, although in principle a parity violation could be detected from phenomenological analyses of processes like π - N scattering small violation would be very hard to pick up, especially in the presence of resonant parity conserving structures.

2.2. TIME REVERSAL

Before considering the quantum theory I shall recall briefly to you time-reversal invariance in classical physics. Classically, T-invariance says simply that to any motion of a system which is permitted by the equations of motion there is another permitted motion, in which the system goes through the same sequence of classical states (points in phase space), but in the opposite order. Here is a list showing the behaviour of some classical dynamical variables when the sense of time is reversed:

TABLE 3

$\vec{r} \to$	$\vec{r}$	
$\vec{v}$	$-\vec{v}$	since each d/dt changes the
$\vec{a}$	$\vec{a}$	sign once.
$\vec{F}$	$\vec{F}$	from the line above, and Newton's equation
$\vec{E}$	$\vec{E}$	from the Lorentz equation, therefore
$\vec{A}$	$-\vec{A}$	from the $\vec{v}$ - dependence, or from $\vec{E} \sim \dot{\vec{A}}$
V	V	from $\vec{E} \sim \vec{\nabla} V$
$\vec{B}$	$-\vec{B}$	from $\vec{B} \sim \vec{\nabla} \times \vec{A}$ or the Lorentz equation
$\vec{J}$	$-\vec{J}$	from $\vec{J} = \rho\vec{v}$, or Maxwell's equation

In discussing the quantum theory, let us consider first the Schrödinger equation in <u>non-relativistic quantum mechanics</u>

$$-\frac{1}{i}\frac{\partial\psi(\vec{x}, t)}{\partial t} = H\psi(\vec{x}, t) \tag{2.16}$$

(we here follow the usual convention in denoting non-relativistic states by <u>small</u> Greek letters).
The complex conjugate of this equation is

$$\frac{1}{i}\frac{\partial\,\psi^*(\vec{x}, t)}{\partial t} = H^*\psi^*(\vec{x}, t).$$

Replacing t by -t throughout this equation gives

$$-\frac{1}{i}\frac{\partial\psi^*(\vec{x}, -t)}{\partial t} = H^*\psi^*(\vec{x}, -t). \tag{2.17}$$

Comparing this with (2.16), we see at once that, provided $H = H^*$, ψ^* may be the same function of -t as ψ is of t.
The* denotes complex conjugation, not hermitian conjugation, and is therefore representation-dependent. We may introduce an operator of complex conjugation K, and write

$$K\,\Psi = \Psi^*$$
$$K(\text{any operator})\,K = (\text{operator})^*. \tag{2.18}$$

with

$$K^2 = 1.$$

In complex-conjugating an operator we simply take the complex conjugate of every matrix element.

Comparing equations (2.16) and (2.17), we see that we do not have time-reversal invariance unless H is real. But if, for example, H depends on spins, then it contains σ_y, which is pure imaginary in the representation (2.5), so that $H^* \neq H$. In order to achieve T-invariance, therefore, we must require that there be a further transformation, which we shall call T_1, which takes H^* into H, so that we have probability amplitudes preserved. For example, spins, if they are present, must be reversed by T_1. In order to preserve scalar products and probability amplitudes, T_1 has to be unitary, and we require it to be time-independent. Its characteristic property is

$$T_1 H^* T_1^{-1} = H \tag{2.19}$$

and in general the time-reversal operator is

$$T = T_1 K. \tag{2.20}$$

It is an anti-unitary, not a unitary, operator.

As an example of the representation-dependence, consider the x-representation of the Schrödinger equation, so that a momentum eigenstate is

$$\psi_{\vec{p}_0}(\vec{x}) = e^{i\vec{p}_0\cdot\vec{x}}$$

and (in the absence of spins)

$$K\,\psi_{\vec{p}_0}(\vec{x}) = e^{-i\vec{p}_0\cdot\vec{x}} = \psi_{-\vec{p}_0}(\vec{x})$$

which is just what we want (in agreement with our table of classical transformation properties), and so we may set

$$T_1 = 1.$$

In the p-representation, on the other hand, the corresponding state is

$$\phi_{\vec{p}_0} = \delta(\vec{p}_0 - \vec{p})$$

which is real, so that

$$K\phi_{\vec{p}_0} = \phi_{\vec{p}_0}$$

which is not the time-reversed state which we want. So we have to do something more: T_1 has to be a unitary operator which changes the sign of $\vec{p}$:

$$T_1 K\phi_{\vec{p}_0}(\vec{p}) T_1^{-1} = \phi_{\vec{p}_0}(-\vec{p}) = \phi_{-\vec{p}_0}(\vec{p})$$

so that the T_1 operator in momentum space has a violent effect on the wave-function, whereas the T_1 operator in coordinate space has no effect whatsoever (in the absence of spin).

Now let us see, still within the non-relativistic framework, what T_1 has to do to spin. Suppose, for example, we have the typical Hamiltonian.

$$H = \vec{\sigma} \cdot \vec{L}.$$

In coordinate space, $\vec{p}$ is pure imaginary, and so is $\vec{L} = \vec{r} \times \vec{p}$, so that

$$\vec{L}^* = -\vec{L},$$

and we need a T_1 which achieves

$$T_1 \vec{\sigma}^* T_1^{-1} = -\vec{\sigma}.$$

In our standard representation (2.5) we may immediately write the solution T_1 of this equation in the form

$$T_1 = \eta_T \sigma_y \qquad (2.21)$$

where η_T is an arbitrary phase factor, not determined by the conditions on T_1.

In general, if we can find an operator $T = T_1 K$ which appropriately transforms the quantum mechanical operators corresponding to the classical variables in Table 3, and which leaves the quantum-mechanical equations of motion of the system unchanged except that it reverses the sign of the time, then we say that time-reversal (T-invariance) holds. If such an operator cannot be found, then time-reversal does not hold.

We now consider an S-matrix formulation of time-reversal, which is really the essence of time-reversal invariance. Without attempting any detailed justification I shall assume that one may write the S-matrix in the form

$$S_{fi} = (\Psi_f^{(-)}, \Psi_i^{(+)}). \tag{2.22}$$

The state $\Psi^{(\pm)}$ is one whose wavepacket may be approximated by a plane wave at $t = \mp \infty$: in coordinate space we have

$$\Psi^{(+)} \approx e^{ip\cdot x} + \frac{1}{r} e^{ipr} f^{(+)}(\theta, \phi)$$
$$\Psi^{(-)} \approx e^{ip\cdot x} + \frac{1}{r} e^{-ipr} f^{(-)}(\theta, \phi),$$

and the i and f just label the different states. <u>If T-invariance holds</u> then $\Psi^{(\mp)}$ is just the time reversal of $\Psi^{(\pm)}$, provided that all the appropriate quantum numbers (momenta, spins, and so on, as in Table 2) are reversed. We formulate this in the assertion

$$T\Psi_a^{(\pm)} = \eta_T \Psi_{\tilde{a}}^{(\mp)} (\text{with } |\eta_T|^2 = 1) \tag{2.23}$$

where $\tilde{a}$ denotes the operation of reversing the quantum numbers. Note that $\Psi_{\tilde{a}}$ may differ from Ψ_a by a phase. The effect of T on an inner product may be calculated from the assumption that it has the form

$$T = T_1 K \tag{2.20}$$

as follows:*

---

*The operation we are discussing is the one called <u>Wigner</u> time reversal in the literature.

$$(T\Phi,\ T\Psi) = (T_1\Phi^*,\ T_1\Psi^*) = (\Phi^*,\ \Psi^*)$$
$$= (\Phi,\ \Psi)^* = (\Psi,\ \Phi). \qquad (2.24)$$

Combining this with (2.23), we find immediately

$$S_{fi} = (\Psi_f^{(-)},\ \Psi_i^{(+)}) = (T\Psi_i^{(+)},\ T\Psi_f^{(-)}) = (\Psi_{\tilde{i}}^{(-)},\ \Psi_{\tilde{f}}^{(+)}) = S_{\tilde{i}\tilde{f}}, \qquad (2.25)$$

so that with our proviso about reversing thė quantum numbers, the S-matrix is "almost" symmetric. In a helicity and angular momentum representation, a two-particle S-matrix is symmetric.

We may express this same result in operator language. We have:

$$(f, T^{-1} ST\, g) = (Tf,\ ST g)^* \quad \text{(from 2.24)}$$
$$= (\tilde{f},\ S\, \tilde{g})^*$$
$$= (g,\ Sf)^* \quad \text{(from 2.25)}$$
$$= (f,\ S^\dagger g)$$

or

$$T^{-1}\, ST = S^\dagger = S^{-1}. \quad (2.27)$$

Exercise: Verify that $T^{-1}\, S^\dagger\, T = S$ (2.27)

A last remark about the theory: since T is not unitary, we cannot be sure a priori that a change of representation in the Hilbert space preserves the form $T = T_1 K$, T_1 unitary. In fact it does; this may be seen as follows: consider the unitary transformation

$$\Psi_n \rightarrow \Psi_n' = U\, \Psi_n$$

(the two sets of states are labelled by the same quantum numbers). We demand that the transformed T have the same effect on the transformed states as the original T on the original states:

$$T'(U\, \Psi_n) = T'\Psi_n' = \Psi_{\tilde{n}}' = U\Psi_{\tilde{n}} = (UT)\Psi_n \quad .$$

If now $T' = T_1' K$, then we must have

$$T_1' KU = UT_1 K$$

or

$$T_1' U^* K = UT_1 K$$

so that

$$T_1' U^* = UT_1$$

or

$$T_1' = UT_1 (U^*)^{-1}$$

which shows (after an easy calculation) that if T_1 is unitary, so is T_1'.

We now consider some applications:

1. <u>Born approximation calculations,</u> especially in weak interactions. We write S in terms of the scattering amplitude matrix*, or transition matrix, M:

$$S = 1 + i\,(2\pi)^4\,\delta^{(4)}(p)\,M. \qquad (2.28)$$

Then the matrix M has the same symmetry properties as S. In the Born approximation, the first non-vanishing transition matrix is normally <u>(always,</u> if it is first order) hermitian. We consider the consequences of hermiticity. It turns out that in this case one can relate a process to itself and not to the time-reversed process, which gives one a very simple way of inspecting formulae for time-reversal invariance.

Inserting (2.28), which defines M, into (2.25), namely $S_{fi} = S_{\tilde{i}\tilde{f}}$, we have at once, by time reversal invariance

$$\langle f|M|i\rangle = \langle\tilde{i}|M|\tilde{f}\rangle = \langle\tilde{f}|M^{\dagger}|\tilde{i}\rangle^* \qquad (2.29)$$

and so <u>if M is Hermitian</u>

---

*What is here denoted M is not really the Moller matrix (an energy-dependent factor has been omitted), but is what is normally denoted "-T matrix;" we write M to avoid a serious confusion of notation.

$$\langle f|M|i\rangle = \langle \tilde{f}|M|\tilde{i}\rangle^*. \tag{2.30}$$

and

$$|\langle f|M|i\rangle|^2 = |\langle \tilde{f}|M|\tilde{i}\rangle|^2 \tag{2.31}$$

This can be applied immediately to processes like β-decay. Can there exist, for example, a correlation like

$$\vec{\sigma} \cdot \vec{p}_1 \times \vec{p}_2 \ ?$$

The answer is no if the system is self-adjoint, because time-reversal changes the signs in all three factors, hence the sign of the whole expression. If the system is not self-adjoint, then this term can of course occur—it is exactly this term in the probability which produces the polarization in π - N scattering, for example. It should be emphasized that the hermiticity is essential for relating a process to itself— in the example of β-decay, the time reversal of $n \rightarrow p + e + \bar{\nu}$ is not itself, but $p + e + \bar{\nu} \rightarrow n$, and only the hermiticity restores the original reaction.

2. Final state interactions. Consider an initial state without interactions, for example

$$\Lambda \rightarrow \pi + N$$

The initial state is essentially real, in the sense that there is no distinction between incoming and outgoing states. However, in a π + N state, there is a distinction. Let us consider a matrix element of the form

$$(\Psi_f^{(-)}, V\Phi_i) = (\text{by time-reversal})\ (\Phi_i, V\Psi_{\tilde{f}}^{(+)}) = (\Psi_{\tilde{f}}^{(+)}, V\Phi_{\tilde{i}})^*,$$

with V self-adjoint. If $\Psi_f^{(-)}$ contains only a few partial waves, then there is a simple relation between $\Psi_f^{(-)}$ and $\Psi_f^{(+)}$.

Assuming that a partial wave analysis is possible (as in the π - N case, where it is completely kinematic), we have, for a definite partial wave, $\Psi_f^{(\pm)}$ has the phase $e^{\pm i\delta}$, so that

$$\Psi_f^{(+)} = e^{2i\delta}\ \Psi_f^{(-)}.$$

Substituting in the matrix element, we find, for this given partial wave,

$$(\Psi_f^{(-)}, V\Phi_i) = (e^{2i\delta}\Psi_{\tilde{f}}^{(-)}, V\Phi_{\tilde{i}})^*.$$

It can be seen from this that we may write the interaction as if it were a hermitian one by setting

$$\Psi_f^{(-)} = \Phi_f e^{-i\delta}$$

for then the relation between the matrix elements becomes

$$(e^{-i\delta}\Phi_f, V\Phi_i) = (e^{i\delta}\Phi_{\tilde{f}}, V\Phi_{\tilde{i}})^*$$

This is useful, for example, in analysing $\Lambda \rightarrow N + \pi$ or $\Sigma \rightarrow N + \pi$, where the M-matrix is not self-adjoint, but only S and P waves are involved and all the non-self-adjointness is contained in the final state phases. Knowing the phases from scattering experiments, one can repeat the calculations, almost as simply as in the hermitian case, and search for violations of time-reversal invariance.

3. Permissible spin dependence of elastic scattering amplitudes. Again we work in a non-relativistic, non-field-theoretic framework, but we shall see that identical results come out of the relativistic theory. We consider the spin matrix element for an essentially elastic process

$$(\chi^*_{f\alpha} M_{\alpha\beta} \chi_{i\beta}) \tag{2.32}$$

(This is a matrix product with the spinor indices α, $\beta = 1, 2$ put in explicitly).

The χ's denote the spin states. We assume that the momentum-dependence of the matrix element is already contained in $M_{\alpha\beta}$. For an elastic process, we assume that the matrix $M_{\alpha\beta}$ describes the inverse process as well as the direct process. The time-inverted matrix element is

$$(\chi^*_{\tilde{i}\alpha} M'_{\alpha\beta} \chi_{\tilde{f}\beta}) \tag{2.33}$$

where

$$M'_{\alpha\beta}(\vec{p}_i\,,\ \vec{p}_f) = M_{\alpha\beta}(-\vec{p}_f\,,\ -\vec{p}_i), \qquad (2.34)$$

and if T-invariance holds, the two matrix elements must be equal. The momentum variables have been inverted directly by the change from $M_{\alpha\beta}$ to $M'_{\alpha\beta}$; to invert the spin variables we make use of (2.20) and (2.21), which give

$$\chi_{\tilde{i}\alpha} = (\eta_T \sigma_y \chi_i^*)_\alpha .$$

Inserting this into (2.33) we obtain

$$\begin{aligned}(\eta_T \sigma_y \chi_i^*)_\alpha^* M'_{\alpha\beta}(\eta_T \sigma_y \chi_f^*)_\beta &= (\sigma_{y\alpha\gamma} \chi_{i\gamma}^*)^* M'_{\alpha\beta}(\sigma_{y\beta\delta} \chi_{f\delta}^*) \\ &= \chi_{i\gamma} \tilde{\sigma}_{y\gamma\alpha}^* M'_{\alpha\beta} \sigma_{y\beta\delta} \chi_{f\delta}^* \\ &= \chi_{i\gamma} \sigma_{y\gamma\alpha} M'_{\alpha\beta} \sigma_{y\beta\delta} \chi_{f\delta}^* .\end{aligned}$$

For T-invariance, this must be equal to (2.32)

$$\begin{aligned}\chi_{f\alpha}^* M_{\alpha\delta} \chi_{i\delta} &= \chi_{i\gamma} \sigma_{y\gamma\alpha} M'_{\alpha\beta} \sigma_{y\beta\delta} \chi_{f\delta}^* \\ &= \chi_{f\alpha}^* \widetilde{(\sigma_{y\alpha\beta} M'_{\beta\gamma} \sigma_{y\gamma\delta})} \chi_{i\delta}\end{aligned}$$

(remember that $\widetilde{\text{(operator)}}$ denotes the transpose, and the spinor indices on the matrices are written in explicitly). Suppressing the spinor indices, we conclude that for T-invariance

$$M = \widetilde{(\sigma_y M' \sigma_y)}. \qquad (2.35)$$

Now let use see what processes are allowed and forbidden on account of this relation. Suppose that M has the form (which you would expect for a single spin $\frac{1}{2}$ particle)

$$M \sim \vec{\sigma} \cdot \vec{A}$$

and let $\vec{A}'$ denote the result of making, in $\vec{A}$, the changes required by (2.34) (namely interchanging the initial and final momenta and changing their signs). From (2.5) a straightforward calculation gives

$$\widetilde{(\sigma_y \vec{\sigma} \sigma_y)} = -\vec{\sigma}.$$

Now for elastic scattering all scalars are T-invariant anyhow, so we need consider only the form of possible vectors $\vec{A}$, which to accord with (2.35) must therefore satisfy

$$\vec{A}' = -\vec{A}.$$

For example, in the center of mass system,

$\vec{A} = \hat{p}_i \times \hat{p}_f \rightarrow \vec{A}' = (-\hat{p}_f) \times (-\hat{p}_i) = -\vec{A}$ allowed by T (also allowed by P)

$\vec{A} = \hat{p}_i + \hat{p}_f \rightarrow \vec{A}' = (-\hat{p}_f) + (-\hat{p}_i) = -\vec{A}$ allowed by T

$\vec{A} = \hat{p}_i - \hat{p}_f \rightarrow \vec{A}' = (-\hat{p}_f) - (-\hat{p}_i) = \vec{A}$ forbidden by T

I should like to make a point here about terminology:

Consider, for example, π - N scattering, for which the scattering amplitude may be written in the form

$$f = f_1 + (\vec{\sigma} \cdot \hat{p}_2)(\vec{\sigma} \cdot \hat{p}_1) f_2. \tag{2.36}$$

If there are no absorptive processes (so that there are no $i\epsilon$'s in the energy denominators) then the Hamiltonian is effectively Hermitian. It is convenient, and customary, to write the scattering amplitude so that the departure from Hermiticity appears entirely in the imaginary parts of the coefficients f_1 and f_2. So if we use the properties of Pauli matrices to rewrite (2.36) in the form

$$f = \underset{\text{non-spin-flip scattering}}{(f_1 + f_2 \cos\theta)} + \underset{\text{spin-flip scattering}}{i\,\vec{\sigma} \cdot \hat{p}_1 \times \hat{p}_2\, f_2}$$

the coefficient of $\vec{\sigma} \cdot \hat{p}_1 \times \hat{p}_2$ contains i, but if f_2 is real we still say that this spin-flip scattering term is real.

We come now to the relativistic theory of time-reversal. We introduce an operator T_{op} in Hilbert space, whose effect on a Dirac field operator is expressed by an equation corresponding to (2.13):

$$T_{op}\,\psi_\alpha(\vec{x},\ t)\ T_{op}^{-1} = \eta_T T_{\alpha\beta}\,\psi_\beta(\vec{x},\ -t) \tag{2.37}$$

Again $T_{\alpha\beta}$ is a numerical matrix which we shall determine and η_T is a phase factor. The operator T_{op} time-reverses field operators and complex conjugates c-numbers but does not take field operators into their adjoints. Matters are arranged so that the complex conjugation required in (2.24) is achieved by transforming the state vectors. As before, we can determine $T_{\alpha\beta}$ by operating on the Dirac equation with T_{op} leaving out the scalar and pseudo-scalar fields (which will be dealt with later) we write the Dirac equation **(2.12) in the** form

$$-\frac{1}{i}\frac{\partial\psi_\alpha(\vec{x},t)}{\partial t} = \left[\rho_1\vec{\sigma}\cdot\left(\frac{\nabla}{i} - e\vec{A}(\vec{x},t)\right) + \beta m\right]\psi(\vec{x},t)$$

and perform T_{op} (equation) T_{op}^{-1}, to get

$$+\frac{1}{i}\frac{\partial}{\partial t}T_{\alpha\beta}\psi_\beta(\vec{x},-t) = \left[\rho_1\vec{\sigma}^*\left(-\frac{\nabla}{i} - e\vec{A}'(\vec{x},t)\right) + \beta m\right]T_{\alpha\beta}\psi_\beta(\vec{x},t)$$

where

$$\vec{A}'(\vec{x},t) = T_{op}\,\vec{A}(\vec{x},t)\ T_{op}^{-1}. \tag{2.38}$$

Taking the $T_{\alpha\beta}$ from left to right, suppressing the spinor indices, and replacing $-t$ by t, we obtain

$$-\frac{1}{i}\frac{\partial}{\partial t}\psi(\vec{x},t) = T^{-1}\left[\rho_1\vec{\sigma}^*\left(-\frac{\nabla}{i} - e\vec{A}'(\vec{x},-t)\right) + \beta m\right]T\psi(\vec{x},\ t)$$

which is the same as the original equation provided that

$$T^{-1}\rho_1\vec{\sigma}^*T = -\rho_1\vec{\sigma}$$
$$T^{-1}\beta T = \beta.$$

Both these equations are satisfied by

$$T = \sigma_y \text{ and } \vec{A}'(\vec{x},\ -t) = -\vec{A}(x,t)$$

so that (2.37) becomes

$$T_{op}\,\psi_\alpha(\vec{x},t)\ T_{op}^{-1} = \eta_T\sigma_{y\,\alpha\beta}\ \psi_\beta(\vec{x},\ -t). \tag{2.39}$$

Exercise: Check that the commutation relation (2.39) is compatable with the transformation of the spinor wave function

$$\chi e^{imt} = (\Psi_{vac}, \psi(\vec{x}, t)\Psi_p),$$

by time reversal, into

$$\eta_T \sigma_y \chi^* e^{-imt} = (T_{op}\Psi_{vac}, \psi(\vec{x}, t) T_{op}\Psi_p) .$$

Solution: From (2.24) we have

$$(T_{op}\Psi_{vac}, \psi(\vec{x}, t) T_{op}\Psi_p) = (\Psi_{vac}, T_{op}^{-1}\psi(\vec{x}, t) T_{op}\Psi_p)^*$$

and so from (2.39), $\quad = (\Psi_{vac}, \eta_T \sigma_y \psi(\vec{x}, -t)\Psi_p)^*$

which gives the required result.

Exercise: Consider the reactions

$$\pi^+ + D \rightarrow p + p \text{ with cross-section} \left(\frac{d\sigma}{d\Omega}\right)_1$$

$$p + p \rightarrow \pi^+ + D \text{ with cross-section} \left(\frac{d\sigma}{d\Omega}\right)_2$$

Calculate the ratio of the cross-sections at any given angle, assuming that there is no measurement of polarization, or preparation of a polarized state.

Solution: (Källén, <u>Elementary Particle Physics</u>, Sec. 3.2): The principle of detailed balance, deduced from time reversal invarance, gives

$$\frac{\left(\frac{d\sigma}{d\Omega}\right)_1}{\left(\frac{d\sigma}{d\Omega}\right)_2} = \frac{4}{3}\,\frac{\lambda(s,\ M_p^2,\ M_p^2)}{\lambda(s,\ m_\pi^2,\ M_d^2)}$$

where the λ's are the usual kinematical factors.

We now discuss the effect on the adjoint $\psi^\dagger$ of the transformation (2.39). In order to do this, we decompose T_{op}, according to (2.20), into a unitary operator and a complex conjugation:

$$T_{op} = T_{1op}\ K.$$

Then (2.39) becomes

$$T_{1op}\ K\,\psi_\alpha(\vec{x},\ t)\cdot KT_{1op}^{-1} = \eta_T \sigma_{y_{\alpha\beta}} \psi_\beta(\vec{x},\ -t)$$

since $K^2 = 1$.

Assuming that complex conjugation of the field operator commutes with taking of the adjoint, we may take the adjoint of this equation, obtaining (since T_{1op} is unitary)

$$T_{1op}\,K\,\psi_\alpha^\dagger\ (\vec{x},\ t)\ KT_{1op}^{-1} = \eta_T^* \,\sigma_{y_{\alpha\beta}}^*\ \psi_\beta^\dagger\ (x,\ -t)$$
$$= -\ \eta_T^*\,\sigma_{y_{\alpha\beta}}\ \psi_\beta^\dagger\ (\vec{x}, -t)$$

since in our standard representation σ_y is pure imaginary. We therefore conclude that

$$T_{op}\,\psi_\alpha^\dagger(\vec{x}, t)\,T_{op}^{-1} = -\ \eta_T^*\,\sigma_{y_{\alpha\beta}}\ \psi_\beta^\dagger\,(\vec{x}, -t) \qquad (2.40)$$

Let us consider the time-reversal properties of some other interactions. Instead of putting the interactions explicitly into the Dirac equation, as in (2.12), we may now, with the help of (2.39) and (2.40), consider the properties of interaction currents directly. For example, consider the pseudo-scalar current

$$H_{ps}' = \psi^\dagger \rho_2 \psi\, \phi_{ps}$$

where ϕ_{ps} is the pseudo-scalar field operator. We have

$$T_{op}\,H_{ps}'\,T_{op}^{-1} = (-\eta^*\sigma_y\,\psi^\dagger)\,(\rho_2^*)\,(\eta\sigma_y\,\psi)\,\phi'$$

where

$$\phi'\,(\vec{x}, -t) = T_{op}\,\phi(\vec{x},\ t)\ T_{op}^{-1}$$
$$= (-\eta^*)\,(\psi^\dagger\tilde{\sigma}_y)\,(-\rho_2)\,(\eta\sigma_y\psi)\,\phi' = -\ \psi^\dagger\rho_2\psi\phi'$$

(since $\eta\eta^* = 1$ and σ_y is antisymmetric). Therefore for T-invariance we must have

$$\phi' = -\phi.$$

It turns out that the pseudo-vector coupling $i\overline{\psi}\gamma_5\gamma_\mu\psi\partial_\mu\phi_{ps}$ gives the <u>same</u> condition on ϕ, but there is no guarantee of consistency in all possible cases. For example, in the case of a scalar field, we may write coupling terms

$$H_s' = g_s\psi^\dagger\beta\psi\phi_s + g_v\psi^\dagger\vec{\alpha}\psi\cdot\vec{\nabla}\phi_s \text{ (+ 4}^{\text{th}}\text{ component)}$$

and in this case the first term, which is like a mass term, is T-invariant, but the second term changes sign, because $\vec{\alpha}$ changes sign. It appears that this coupling leads to a non-T-invariant theory. However, it happens that in this case the derivative coupling in the second term really leaves the system effectively uncoupled, so that this example does not lead to an inconsistency. For the general case, I can state the following <u>Rule</u> (to be derived later): Yukawa couplings of neutral <u>tensor</u> fields (spin J and parity $(-1)^J$) to spin $\frac{1}{2}$ or spin zero can always be put into T-invariant form; this is not true for <u>pseudo-tensor</u> fields (spin J and parity $-(-1)^J$), except for $J = 0$.

The first case which arises is the subject of the

Exercise: Show that T-invariance can be violated by a spin $\frac{1}{2}$ field coupled to an axial vector field.

We now discuss the <u>experimental evidence for time-reversal invariance</u>. The evidence from $\pi^+ + D \leftrightarrow p + p$, which was the original basis for determination of the pion spin (problem 2), is not very strong evidence for T-invariance. Neither is the evidence from β-decay. Among the strong interactions, however, there are many cases in which one can determine the cross-sections for an experiment ($a + b \rightarrow c + d + \ldots$) and the reciprocal experiment ($c + d + \ldots \rightarrow a + b$), and verify that the cross-sections are appropriately related. There is also some evidence from N - N scattering. The decay $K_2 \rightarrow 2\pi$ implies a violation of CP, so that if CPT holds,

T <u>is</u> violated at least in the weak interactions, if not the strong. On the other hand, in nucleon physics, so long as isotopic spin is conserved, it is impossible (by the <u>Rule</u>) to violate T until we reach at least the complication of psuedo vector forces, or virtual n-particle processes. Therefore, a strong violation of T has a strong tendency to be concealed. This will be discussed in more detail later, in connection with charge conjugation.

2.3. CHARGE CONJUGATION

The operation of charge conjugation, C, is essentially Hermitian conjugation. This arises because in the interpretation of the field theory, an operator ψ, for example, annihilates an electron or creates a positron, and so decreases the charge by e, while the adjoint operator $\psi^\dagger$ creates an electron or annihilates a positron, and so increases the charge by e. Naturally, we want C, if it is to change the signs of all charges, to take either of these operators into the other.

Accordingly, we seek an operator C_{op} on the Hilbert space with the property

$$C_{op}\,\psi_\alpha(\vec{x}, t)\,C_{op}^{-1} = C_{\alpha\beta}\,\psi_\beta^\dagger\,(\vec{x}, t) = \psi'_\alpha\,(\vec{x}, t). \qquad (2.41)$$

(compare equations (2.13) and (2.37)). Suppressing the indices on the numerical matrix $C_{\alpha\beta}$ and on ψ_β, we have

$$C_{op}\,\psi(\vec{x}, t)\;C_{op}^{-1} = C\,\psi^\dagger\,(\vec{x}, t). \qquad (2.41')$$

We shall try to transform the Dirac equation in such a way as to be consistent with

$$\frac{\partial C_{op}}{\partial t} = 0.$$

Sometimes one makes use of a related numerical matrix C_1, which is defined by

$$C_{op}\,\psi\,C_{op}^{-1} = C_1\overline{\psi}.$$

Comparing this with (2.41'), and remembering that $\bar{\psi} = \psi^\dagger \beta^\dagger = \tilde{\beta}\psi^\dagger = \beta\psi^\dagger$, we have

$$C_1 \beta \psi^\dagger = C\psi^\dagger$$

so that

$$C_1 = C\beta. \tag{2.42}$$

(Here, as always, we work in our standard representation). As we shall see, the matrix C_1 has simpler invariance properties than C.

Now let us transform the Dirac equation (2.12) with C_{op}. We start with

$$-\frac{1}{i}\frac{\partial}{\partial t}\psi(\vec{x}, t) = \left[\vec{\alpha}\cdot\left(\frac{\vec{\nabla}}{i} - e\vec{A}\right) + \beta m + eV\right]\psi(\vec{x}, t) \tag{2.43}$$

The Hermitian conjugate of this equation is

$$\frac{1}{i}\frac{\partial}{\partial t}\psi^\dagger(\vec{x},t) = \left[\vec{\alpha}^*\cdot\left(-\frac{\vec{\nabla}}{i} - e\vec{A}\right) + \beta^* m + eV\right]\psi^\dagger(\vec{x},t) \tag{2.44}$$

Strictly speaking we should interchange the order of the operator ψ and $\vec{A}$, V, but in the absence of derivative coupling this makes no difference, since at equal times these operators commute. Also $\vec{A}$ and V are self-adjoint, and therefore unchanged by the conjugation. All the c-numbers, on the other hand, must be replaced by their complex conjugates. To preserve the commutation rules for ψ, C_{op} must be a unitary transformation.

Problem: Show that if C_{op} is a unitary operator then C is a unitary matrix:

$$C_{\alpha\beta}\, C^*_{\alpha\delta} = \delta_{\beta\delta}. \tag{2.45}$$

Solution: Use the anti-commutation relations for ψ_α,

$$\{\psi_\alpha(\vec{x}), \bar{\psi}_\beta(\vec{x}')\} = i\,\delta_{\alpha\beta}\,\delta(x - x') \tag{2.46}$$

Now let us charge conjugate the Dirac equation (2.43) by calculating C_{op} (equation) C_{op}^{-1}. From (2.41) we find

$$-\frac{1}{i}\frac{\partial}{\partial t} C\,\psi^\dagger = \left[\vec{\alpha}\cdot\left(\frac{\vec{\nabla}}{i} - e\,\vec{A}'\right) + \beta m + eV'\right] C\,\psi^\dagger$$

where

$$\vec{A}' = C_{op}\,\vec{A}\,C_{op}^{-1}$$
$$V' = C_{op}\,V\,C_{op}^{-1}.$$

Multiplying by $-C^{-1}$ gives

$$\frac{1}{i}\frac{\partial}{\partial t}\psi^\dagger = C^{-1}\left[\vec{\alpha}\cdot\left(-\frac{\vec{\nabla}}{i} + e\vec{A}'\right) - \beta m - eV'\right] C\psi^\dagger_\beta .$$

Identifying this with the Hermitian conjugate equation (2.44) we require

$$C^{-1}\vec{\alpha}C = \vec{\alpha}^*$$
$$C^{-1}\beta C = -\beta$$
$$\vec{A}' = -\vec{A}$$
$$V' = -V$$

(As one would expect, the 4-potential changes sign under charge conjugation). These equations determine C up to a phase--it is easily seen to be

$$C = \eta_c \rho_2 \sigma_y. \qquad (2.47)$$

Following our established pattern, let us calculate the transformation properties of the currents listed in Table 2 (p. 490) under charge conjugation. Let

$$J = \bar{\psi}\,\Gamma\,\psi,$$

where Γ is any of the invariant tensors appearing in the currents in Table 2.

We have at our disposal the following relations:

$$\left.\begin{aligned}
&\beta = \rho_3 = \tilde{\beta} = \beta^* = \beta^\dagger, \ \beta^2 = 1. \\
&C = \eta_c \rho_2 \sigma_y = \tilde{C} \\
&CC^\dagger = 1 \\
&C_1 = C\beta, \text{ so that } \tilde{C}_1 = \tilde{\beta}\tilde{C} = \beta C = -C\beta = -C_1, \text{ and} \\
&C_1 C_1^\dagger = C\beta\beta^\dagger C^\dagger = CC^\dagger = 1. \\
&\bar{\psi} = \beta\psi^\dagger \\
&C_{op}\,\psi\, C_{op}^{-1} = C\psi^\dagger = C\beta^2\,\psi^\dagger = (C\beta)(\beta\psi^\dagger) = C_1\bar{\psi} \\
&C_{op}\,\bar{\psi}\, C_{op}^{-1} = \beta C^*\psi = -C^*\beta\,\psi = -C_1^*\psi = -\psi C_1^\dagger
\end{aligned}\right\} \quad (2.48)$$

With the help of these, we find very easily that

$$\begin{aligned}
C_{op}\ J\, C_{op}^{-1} &= C_{op}\,\bar{\psi}\, C_{op}^{-1}\ \Gamma\ C_{op}\ \psi\, C_{op}^{-1} = -\psi C_1^\dagger\ \Gamma\, C_1 \bar{\psi} \\
&= \bar{\psi}\ C_1\ \tilde{\Gamma}\ C_1^{-1}\ \psi.
\end{aligned}$$

Here we have used also the anticommutation relations for ψ (we may either ignore the δ-function in (2.46), or consider the normal-ordered current operator). The rule for transforming currents is therefore determined by

$$\Gamma \rightarrow C_1\ \tilde{\Gamma}\ C_1^{-1}. \qquad (2.49)$$

The currents and their transformations are listed in Table 4.

TABLE 4.

	Current	Effect of C (invariant: +, or changes sign: -)
S:	$\bar{\psi}\,\psi$	+
V:	$\bar{\psi}\,\gamma^\mu\psi$	-
T:	$\bar{\psi}[\gamma^\mu, \gamma^\nu]\psi$	-
A:	$\bar{\psi}\gamma_5\gamma^\mu\psi$	+
P:	$\bar{\psi}\,\gamma_5\,\psi$	+

The easily verified results $C_1\tilde{\gamma}_\mu C_1^{-1} = -\gamma_\mu$, $C_1\tilde{\gamma}_5 C_1^{-1} = \gamma_5$, are used in constructing the table.

It follows, for example, the combination of pseudo-scalar and derivative coupling of a pseudo-scalar meson described by

$$\gamma_5\,\phi_{ps} + \gamma_5\,\gamma_\mu\,\partial_\mu\,\phi_{ps}$$

is invariant under charge conjugation, while the combination of scalar and derivative coupling

$$\phi_s + \gamma_\mu\,\partial_\mu\,\phi_s$$

is not, since the first term is invariant, while the second term changes sign under C. However, as in the case discussed on p.505, the second term can be transformed away.

Exercise: Verify that $[\vec{J}, C_{op}] = 0$. [For spin - $\frac{1}{2}$ fields, recall that

$$\vec{J} = \int d^3x\ \psi^+(\vec{x})\left(\frac{\vec{\sigma}}{2} + \vec{r}\times\frac{\vec{\nabla}}{i}\right)\ \psi(\vec{x})].$$

This implies that under rotations ψ' transforms in the same way as ψ (while on the other hand ψ^+ transforms under the complex conjugate representation)].

Exercise: Examine the commutation properties of P_{op} and C_{op} for spin $\frac{1}{2}$ particles

a) construct $P_{op} = \exp(\text{something})$, including the possibility of an arbitrary phase for η_p.

b) carry out $C_{op}\,P_{op}\,C_{op}^{-1}$, and choose the phase so that $[C_{op}, P_{op}] = 0$.

[The phase turns out to be imaginary, which corresponds to the intrinsically negative parity of a particle-anti-particle system].

Remark: P_{op} for system of interacting field = $\Pi\ P_{op}$'s for each separately

Exercise: Determine ψ' when ψ is expanded in momentum and spin states:

$$\psi(\vec{x}) = \sum_{\vec{p},s} A_{\vec{p},s} U_{\vec{p},s} e^{i\vec{p}\cdot\vec{x}} + \sum_{\vec{p},s} b^{+}_{\vec{p},s} (\text{factor to be found}) U^{*}_{p,s} e^{-i\vec{p}\cdot\vec{x}}$$

Outline solution: ψ' will have the form

$$\psi'(\vec{x}) = \sum_{\vec{p},s} b_{\vec{p},s} U_{\vec{p},s} e^{i\vec{p}\cdot\vec{x}} + \sum_{\vec{p},s} (\text{terms of form to be found})$$

In the calculation, set $\eta_c = 1$, and make use of

$$\psi^{\dagger} = C^{\dagger}\psi',$$
$$\psi = C\psi'^{\dagger}.$$

The missing factor in the second term of ψ is found from these equations to be simply C. The term C^*u^* satisfies the equation

$$(-\vec{\alpha}\cdot\vec{p} + \beta m)\, C^*u^* = -EC^*u^*,$$

as we see by writing the left hand side in the form

$$[(-\vec{\alpha}^*\cdot\vec{p} + \beta^* m)Cu]^* = -[C(\vec{\alpha}\cdot\vec{p} + \beta m)u]^* = -C^*E^*u^*,$$

because u satisfies the Dirac equation. Therefore the coefficient of $b^{\dagger}$ corresponds to a negative energy spinor with reversed momentum and spin.

Let us next investigate the charge conjugation properties of a system consisting of a single baryon and its <u>own</u> anti-particle, say $N\bar{N}$, coupled to a boson field. The states available to the $N\bar{N}$ system may be characterized by definite values of parity P and angular momentum $\vec{J}$. In the CM system, the wave function may be constructed from the operator

$$\psi'^{\dagger}_{\beta}(-\vec{p})\,\psi^{\dagger}_{\alpha}(\vec{p})\; Y_{\alpha\beta L}^{JM_J}(\Omega_p) \tag{2.50}$$

where $\psi^{\dagger}_{\alpha}(\vec{p})$ is the particle creation operator, $\psi'^{\dagger}_{\beta}(-\vec{p})$ is the anti-particle creation operator and the angle-dependent matrix Y couples them to a state with total angular momentum defined by J, M_J and orbital angular momentum L, which is not unique; its possible values are

$$\left.\begin{array}{ll} L = \left\{\begin{array}{l} J+1 \\ J \\ J-1 \end{array}\right. & \text{for triplet state } (S = 1) \\ \\ L = J & \text{for singlet state } (S = 0). \end{array}\right\} \tag{2.51}$$

We wish to determine the charge conjugation properties for given J. First we determine P. It does not depend on the spin state, only on the value of L. Normally:

$$P = (-1)^L.$$

However, if you consider an anti-particle and a particle together, you must regard the intrinsic parity of the antiparticle as opposite to that of the particle (see exercise on p. 510). Hence the parity of the state is particle-antiparticle pair:

$$P = (-1)^{L+1} \tag{2.52}$$

As for charge conjugation, we note first of all that $C_{op}\,\psi\,C_{op}^{-1} = \psi'$ and $C_{op}\,\psi'\,C_{op}^{-1} = \psi$, which is to say that C interchanges ψ and ψ'. To relate the charge conjugate state to the original state we must therefore restore the original order and invert all the variables. Thus we have, first,

From anticommutation of the operators: (-1).

Now the triplet spin state is symmetric, the singlet state antisymmetric, which brings in a factor.

From exchange of spin states: $(-1)^{S+1}$.

Finally, since the orbital wavefuctions are even (odd) for even (odd) L, we have

From orbital angular momentum: $(-1)^L$.

The product of these factors gives

$$C = (-1)^{S+L}. \tag{2.53}$$

For higher spin fermions there are more possibilities and this argument does not work.

Now let us assume that space parity is conserved, and consider the coupled boson field. We may distinguish two cases:

$$\begin{aligned} &\text{Tensor field:} && P = (-1)^{J} \\ &\text{Pseudo-tensor field:} && P = (-1)^{J+1} \end{aligned} \tag{2.54}$$

Comparing the tensor field, first, with the particle-antiparticle pair we have from (2.52)

$$J = L \pm 1,$$

and then (2.51) tells us that we must have the triplet state:

$$S = 1.$$

Substituting in (2.53), we get <u>uniquely</u>

$$C = (-1)^{L+1} = (-1)^{J} \tag{2.55}$$

So that a spin $\frac{1}{2}$ particle-antiparticle system coupled to a neutral tensor field <u>cannot</u> violate charge-conjugation invariance, if we simply define $C = (-1)^J$ for all neutral tensor fields.

In the case of a pseudo-tensor field, we have by the same argument

$$J = L$$

but now (2.51) offers two possibilities:

$$\begin{aligned} &\text{triplet: } S = 1, && C = (-1)^{J} \\ &\text{singlet: } S = 0, && C = (-1)^{J+1}. \end{aligned}$$

There are therefore two possible charge conjugation states, with the one exception $J = 0$, in which case, of course, the triplet state is excluded. One often finds that low angular momentum states have especially simple properties like this. The first ambiguous case is therefore the axial vector case, and the first interesting C-violation which can be constructed without charged particles arises in just this way, from vector and tensor coupling of an axial vector field. The simplest example of T-violation, with one kind of particle only, arises in the same way. When many kinds of particle are involved, these results no longer hold.

The validity of the above argument depends, moreover, on the occurrence of an $N\bar{N}$ boson vertex. There is nothing to forbid the existence of other particles which do not have such vertices, but no one has seen any evidence for such an unusual particle, for example, a neutral pseudo-scalar particle with C = -1, and no nucleon-nucleon vertex.

This could be formed in the way described by the following diagram

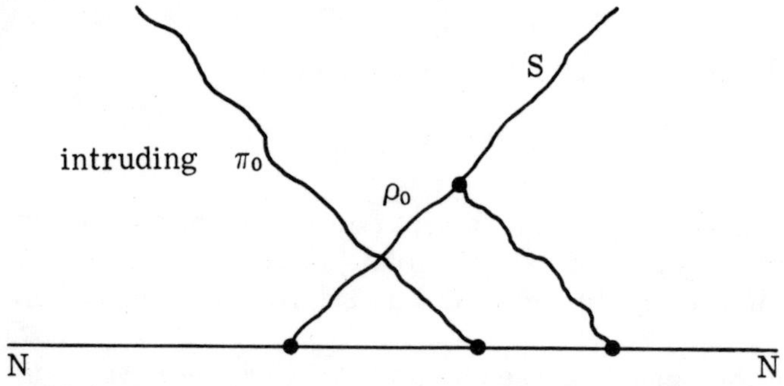

With this rather complicated diagram one could conserve C with a particle having the "wrong" quantum numbers.

The corresponding calculations for bosons are even simpler then those given above. From a spin zero particle and antiparticle one can only construct tensor fields; the pseudo-tensor fields cannot occur. For example the parity of $\pi^+\pi^-$ or K^+K^- is necessarily $P = (-1)^J$. One still finds $C = (-1)^{S+L}$ (but now S = 0) and from J = L concludes that $C = (-1)^J$. For two vector particles, on the other hand, the argument again breaks down.

We postpone the discussion of the validity of charge

conjugation until after an examination of isotopic spin and G-parity, which seem quite critical for this purpose. For the same reason, the discussion of TCP has been postponed until section 3.3 although it should logically have come next.

3. INTERNAL SYMMETRIES AND APPLICATIONS

In this section we shall discuss isotopic spin, G-parity, the experimental verification of C, TCP, and SU_3.

3.1. ISOTOPIC SPIN AND CHARGE INDEPENDENCE

It is remarkable that although we know that I-spin is not an exact quantum number, the experimental support for its conservation is considerably stronger than that for conservation of T or C. The overwhelming evidence is the existence of multiplets with small mass intervals.

Before the discovery of the pion in 1947, the known multiplet was the p - n doublet. It was known that the masses of these particles were very close together; in fact $m_n = 939.55$ MeV, $m_n - m_p = 1.29$ MeV. Charge symmetry between neutron and proton had been adduced from this. In the case of the pion, $m_{\pi^\pm} = 139.58$ MeV, and $m_{\pi^\pm} - m_{\pi^0} = 4.61$ MeV, which is still small, although relatively much larger.

As was suggested by Sakurai,[10] if we neglect the mass differences and assume that the masses are equal, then this very equality implies the existence of a conservation law. Let us see how this comes about, for example, in the case of charge symmetry. Without assuming the isotopic spin formalism suppose that

$$m_p = m_n , \qquad m_{\pi^\pm} = m_{\pi^0}$$

We assume that in zeroth order the masses are equal and look for perturbations which will change them (of course if you assume isotopic spin formalism the equality of the masses is guaranteed to all orders). The first order self-energy diagrams are for δm_p:

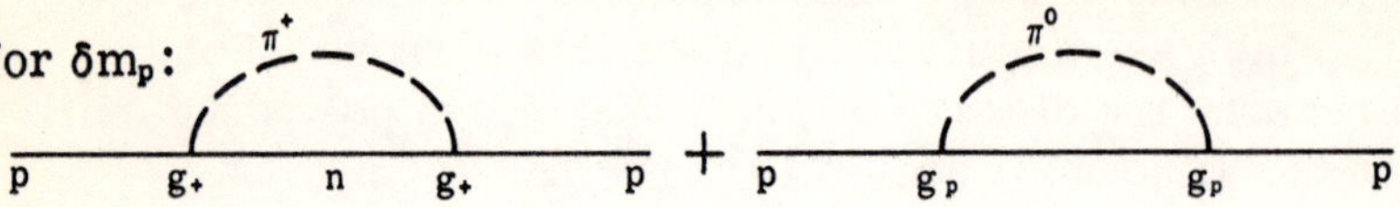

for δm_n:

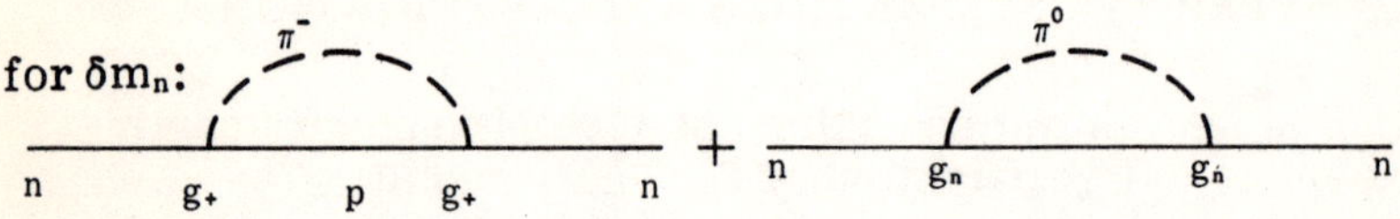

Here the relation between the coupling constants to the charged π's follows from TCP. The requirement of equal masses now implies that

$$g_p^2 = g_n^2 \tag{3.1}$$

but does not tell us whether $g_p = g_n$ or $g_p = -g_n$.

For the pion we can write down the diagrams

π^+:

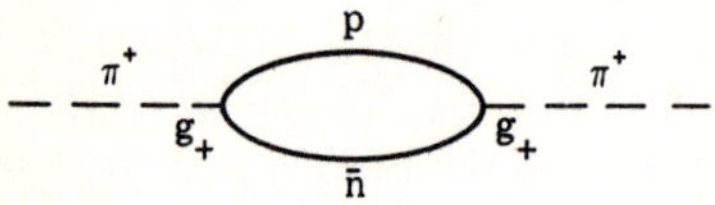

π^0:

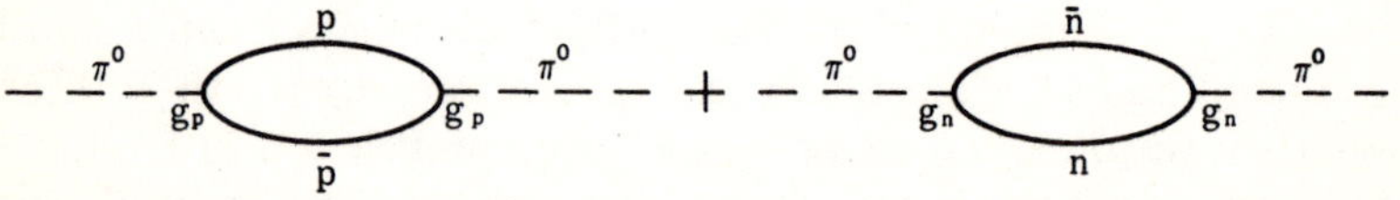

Since we already have (3.1) we can now conclude that

$$\delta\mu_+^2 \propto g_+^2 = 2g_p^2\,. \tag{3.2}$$

To distinguish between the possible signs in (3.1) we go to one higher order. For π^+ there is only basically different fourth-order diagram, namely

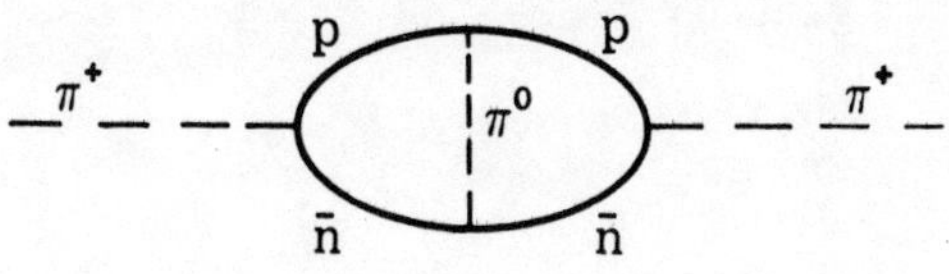

(we exclude the exchange of other particles besides pions). From this diagram we can write down

$$\delta\mu_+^2 \propto g_p\, g_n\, g_+^2.$$

For the π^0 there are many possibilities:

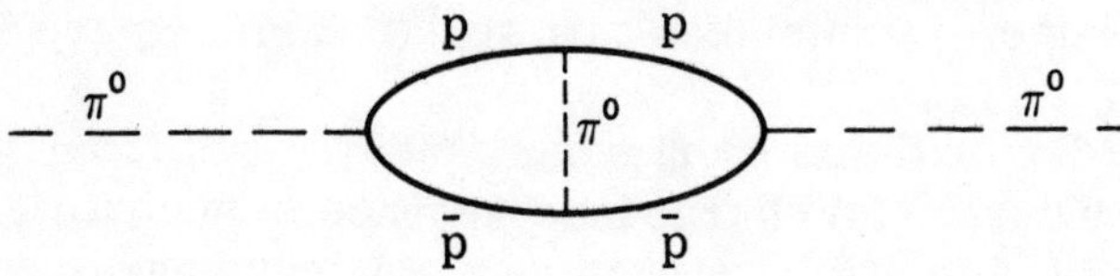

$$\delta\mu_0^2 \propto g_p^4$$

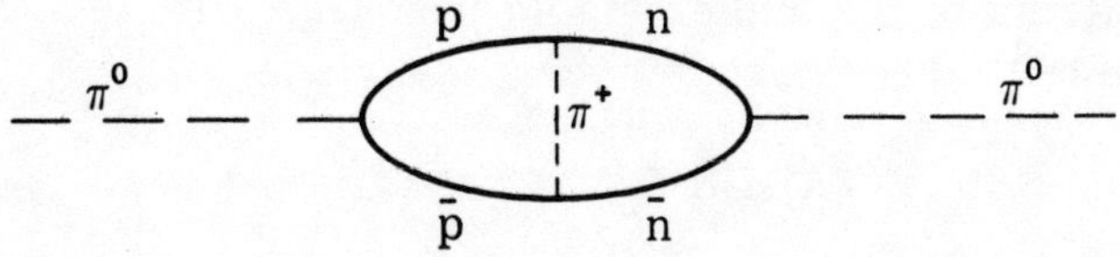

$$\delta\mu_0^2 \propto g_+^2 g_p\, g_n$$

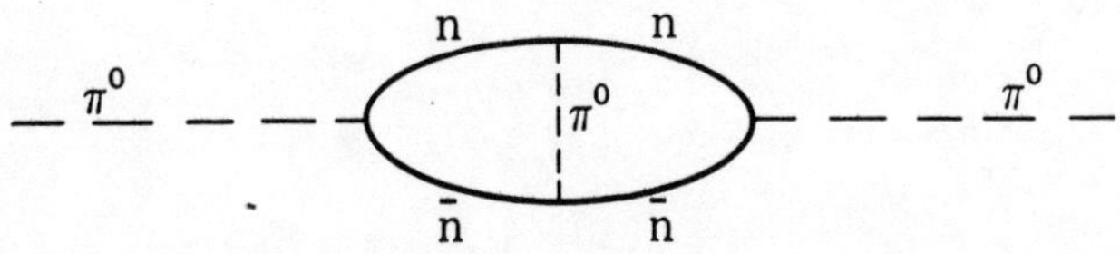

$$\delta\mu_0^2 \propto g_n^4 = g_p^4$$

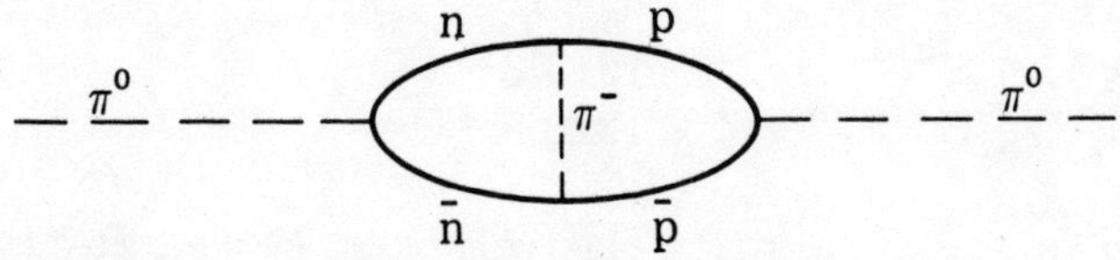

$$\delta\mu_0^2 \propto g_n\, g_p\, g_+^2.$$

Adding these contributions together (and assuming that the masses are still equal to fourth order) we find

$$g_p g_n g_+^2 = 2g_p g_n g_+^2 + 2g_p^4$$

$$2g_p^3 g_n = 4g_p^3 g_n + 2g_p^4$$

whence

$$g_p = - g_n \tag{3.3}$$

which is precisely what we need for charge independence.

Exercise: Using scalar pions and nucleons (for simplicity) prove the charge independence of 1-pion exchange; that is, prove that in a definite state of L, the pp, pn and nn interactions are the same. (Remember that scalar particles are bosons!)

Let us consider further experimental data. Apart from the considerable evidence for charge independence in nuclear interactions from the multiplet structure of nuclear energy levels, there is further evidence from the multiplet structure in the spectrum of the higher stable particles. Table 5 lists their masses.[11]

TABLE 5

Particle	Mass MeV
$K^{\pm}$	493.8
K^0	497.7
Σ^+	1189.4
Σ^0	1192.3
Σ^-	1197.2
Ξ^0	1314.3
Ξ^-	1320.8

We see that the masses are still relatively very close, although the signs of the differences are not the same as for the nucleons. All this is very strong evidence for the existence of the approximate quantum number $\vec{I}$.

But does it make any sense to construct a theory in

which this quantum number is exactly conserved, and try to calculate deviations from this ? We make the implicit assumption that this does make sense—that there is a hypothetical world in which electromagnetic and weak interactions are turned off, and $\vec{I}$ and Y are exactly conserved. This assumption is, in the deepest sense, probably wrong, but we make it anyhow in order to be able to make progress. Phenomenologically, it can be justified by the smallness of the electromagnetic coupling. We make the further hypothesis that the breakdown of charge independence ($\vec{I}$ conservation) arises from electromagnetic interactions, (coupling to photons) and the breakdown of hypercharge conservation arises from weak interactions (coupling to W?) It is a much more difficult question, whether the mass differences are in a serious way accountable to a second order electromagnetic correction to charge independence, although attempts have been made in this direction.[12] But if this is the case, then it follows that for all known particles (for which charge and isotopic spin are related uniquely by an equation of the form

$$Q = "I_3" \text{(isotopic vector)} + "1" \text{(isotopic scalar))} \qquad (3.4)$$

we have

$$Q^2 = A"I_3^2" + B"I_3" + C"1"$$

Therefore to 2nd order in e there must be a sum rule for electromagnetic mass differences of the form

$$\delta m = A\ 1 + B\ I_3 + C\ I_3^2 ,$$

but it has proved very difficult to check such a rule experimentally, because four particles are needed to check a formula with three parameters, which means $I \gtrsim 3/2$. There are no stable particles with $\vec{I}$ large enough, and it is very hard to get the masses of the available unstable particles, namely $N^*3/2$, sufficiently accurately. The formula has been checked for nuclear levels in nuclei with A = 9, and is believed to work very well. However, this is not so surprising, because it is also believed that the main mass differences in nuclei, which are relatively lightly bound, come from the proton-neutron mass differences and the coulomb forces. There are analogous same rules in SU_3, but they check a combination of this electromagnetic violation with SU_3 symmetry. However, they do work excellently there.

There have been attempts to calculate the p - n mass difference, most recently a very interesting one by Dashen and Frautschi[12] in the language of dispersion theory; they consider the proton as a bound state of a pion and a proton, and calculate the effect of an exchange of photons on the binding energy. They get the right sign and the right numerical answer.

We have seen that the near-degeneracy of the nucleon masses and of the pion masses is a strong hint for the existence of a quantum number. However, fifteen years ago people were not so accustomed to thinking of dynamically introduced masses, and also the experimental evidence was much less clear. A crucial experiment carried out in 1951[13] compared the cross-sections for the reactions

$$\begin{array}{ccccc} I=1 & & I=1 & & I=0 \\ \downarrow & & \downarrow & & \downarrow \\ p+p & \rightarrow & \pi^{+} & + & D \\ p+n & \rightarrow & \pi^{0} & + & D \\ \uparrow & & \uparrow & & \uparrow \\ I=1 \text{ or } 0 & & I=1 & & I=0 \end{array}$$

The theoretical cross-sections differ by a factor 2, if I-spin is conserved, and the experiment confirmed this, which is very strong evidence.

Another point I should make is that photon-free reactions which violate charge independence are generally expected go more <u>slowly</u> than photon-containing reactions, because the matrix element for photon production involves e, while that for violation without photons involves e^2, so that the corresponding transition rates go as $e^2 \sim 1/137$ and $e^4 \sim 1/(137)^2$ respectively.

Finally, I should make a remark about <u>currents</u>. Just as the charge (in (3.4)) can be decomposed into an isotopic vector and an isotopic scalar, so can the current:

$$j_\mu = \underset{\text{isovector}}{j_\mu^{(3)}} + \underset{\text{isoscalar}}{j_\mu^{(3)}} . \qquad (3.5)$$

It follows that form factors, magnetic moments, and so on may be expanded into two isotopic terms. This gives little information for the n - p system, because there are only two states, nor for the pion system, because on account of CPT the π^0 has no electromagnetic properties in the one photon approximation. However it does give a relation between the magnetic moments of the three Σ states. This relation is very hard to verify because of the difficulty of measuring the Σ^0 magnetic moment.

3.2. G-PARITY.

This is a convenient aternative to C in the presence of isotopic spin first invented by Michel[14] and called by him "isotopic parity."

A reason for introducing G-parity in addition to its use as a selection rule is essentially that C does not, and must not, commute with isotopic spin:

$$[C, \vec{I}] \neq 0.$$

Why is this? Consider the effect of charge conjugation on a pion field. In order that I_3 be the **charge operator** we must clearly have $I_3 \to -I_3$. More formally:

$$I_3 = \int d^3x\, (\phi_1(x)\Pi_2(x) - \phi_2(x)\Pi_1(x) \qquad (3.6)$$

where $\Pi_i = \dot{\phi}_i$ are the momenta of the field; this can be completed to a vector in I-space by defining

$$\vec{I} = \int d^3x\, \vec{\phi}(x) \times \vec{\Pi}(x). \qquad (3.7)$$

But under charge conjugation we have to exchange positive and negative pions; with a phase factor chosen so that C_{op} is unitary, we have

$$\phi'_+ = C_{op}\ \phi_+\ C_{op}^{-1} = \phi_- = \phi_1 - i\phi_2,$$

$$\phi'_- = C_{op}\ \phi_-\ C_{op}^{-1} = \phi_+ = \phi_1 + i\phi_2,$$

and
$$\phi'_3 = C_{op}\ \phi\ C_{op}^{-1} = \phi_3,$$

so that

$$\begin{aligned} C_{op}\ \phi_1\ C_{op}^{-1} &= \phi_1 \\ C_{op}\ \phi_2\ C_{op}^{-1} &= -\phi_2 \\ C_{op}\ \phi_3\ C_{op}^{-1} &= \phi_3. \end{aligned} \tag{3.8}$$

Inserting these into (3.5) we find

$$\begin{aligned} C_{op}\ I_1\ C_{op}^{-1} &= -\ I_1 \\ C_{op}\ I_2\ C_{op}^{-1} &= \ I_2 \\ C_{op}\ I_3\ C_{op}^{-1} &= -\ I_3 \end{aligned}$$

which shows that C_{op} amounts to a rotation through π around the 2-axis in I-space. Michel suggested cancelling this with a further identical rotation by defining

$$G_{op} = e^{\mp i\pi I_2}\ C_{op} \tag{3.9}$$

which has the property

$$G_{op}\ \vec{I}\ G_{op}^{-1} = \vec{I}.$$

The sign in the exponential makes a difference only for isofermions, for which G_{op} is anyhow not a very useful transformation.

Since G_{op} produces charge conjugation without changing isospin, it has the effects

$$\begin{aligned} G_{op}:\ p &\to \bar{n}\ \ (\text{not } \bar{p}) \\ n &\to -\bar{p}\ \ (\text{not } \bar{n}). \end{aligned}$$

The great beauty of G_{op}, as can be seen from (3.8) and (3.9), is that

$$G_{op}\ \vec{\phi}\ G_{op}^{-1} = -\vec{\phi}$$

$$G_{op}\ \vec{\Pi}\ G_{op}^{-1} = -\vec{\Pi}$$

so that the π-meson has definite G-parity -1 and a state of n π-meson has G-parity $(-1)^n$. In the absence of K's, N's and other particles which do not have definite G-parity, this gives immediately the selection rule

$$\Delta\, n(\pi) = \text{even},$$

whenever charge conjugation and isotopic spin conservation hold.

Let us examine the G-parity of a multiplet which goes into itself under C; it must have N = Y = 0. Then the neutral member of the multiplet will go into itself, say

$$C_{op}\ n = \lambda n \qquad\qquad \lambda = \pm 1$$

where n is the wavefunction for this neutral particle. Since this particle has $I_3 = 0$, we may write the wavefunction in the form

$$n \propto P_I(z)$$

(where z is the 3-coordinate in I-space, and P_I is a Legendre polynomial).
From the definition of G_{op} we now have

$$G_{op}\ n = \lambda\ e^{\pm i\pi I_2}\ P_I(z)$$

$$= \lambda\ P_I(-z)$$

$$= (-1)^I\ \lambda n$$

but since the multiplet as a whole is invariant under G_{op}, we conclude that the G-parity of the entire multiplet is $\lambda\,(-1)^I$.
We tabulate some examples:

Particle	Argument	$G = (-1)\ \lambda$
π	π_0 has $\lambda = 1$(from $\pi \to 2\gamma$) and π is isotopic triplet $\Rightarrow I = 1$.	-1
ρ	Normal vector particle$\Rightarrow \lambda = -1$, isotopic triplet$\Rightarrow I = 1$	+1
ω*	$\lambda = -1$ ($\omega \to 3\Pi$) isotopic singlet$\Rightarrow I = 0$	-1

*Here the argument is backwards: we deduce C from I and G!

Finally, a remark about the G-parity of the photon: To first order, we have the electromagnetic current separated into isoscalar and isovector parts in (3.5). Assuming charge conjugation invariance, we can calculate

$$
\begin{aligned}
G_{op}\, j_\mu G_{op}^{-1} &= e^{i\pi I_2}\ C_{op}\,(j_\mu^{(3)} + j_\mu^{(0)})\,C_{op}^{-1} e^{-i\pi I_2} \\
&= e^{i\pi I_2}\ (-j_\mu^{(3)} - j_\mu^{(0)})\,e^{-i\pi I_2} \\
&= j_\mu^{(3)} - j_\mu^{(0)},
\end{aligned}
$$

so that the isotopic scalar current does change G, but the isotopic vector current does not. Therefore in a linear electromagnetic transition, that is, for single photon emission (real or virtual) the photon changes the G-parity if it carries off $I = 0$, but not if it carries off $I = 1$. Thus the mnemonic: a photon is like a ρ and an ω (or a ϕ).

Exercise: We have seen (p. 513) that $N\bar{N}$ coupled to a pseudoscalar boson has definite C, namely $C = +1$. But what about $\pi^+ \bar{p}\, n$(+h.c.)?

a) Show that π's and N's, with Yukawa coupling, conserve C (irrespective of I-spin conservation).

b) Show that π's, N's, Σ's and Λ's, with Yukawa coupling need not conserve C (assuming only strangeness conservation).

c) Show that π's, N's, Σ's and Λ's conserve C if they conserve I (because there are now enough extra conditions to make the coupling real), but that if K's are added, C need no longer be conserved.

You can see from the results of the exercise that in nuclear physics and pion physics there is a conspiracy of the existing particles to conceal C-violations. This has recently been pointed out very forcefully by Cabibbo.[15]. Incidentally, in systems with nucleon number $\neq 0$, one would effectively have to seek violations of C, assuming CPT, primarily as violations of T.

There is a similar conspiracy among the bosons, but it becomes much less effective if I-spin is not conserved. It is natural to look for C-violation in states which are approximately eigenstates of C, such as neutral bosons. Consider, for example, the ω meson:

$$\omega \to 3\pi$$

Dividing the 3π's into

$$\pi^0 + (\pi + \pi)$$

we observe that I = 1 for the single π, hence I = 1 for the pair, which is therefore in an odd C-state, and C is unique for this system. This also rules out

$$\omega \to \pi^0 + \pi^0 + \pi^0$$

even without explicit C conservation. Now consider the space parity. The ω and the π's each have odd intrinsic parity, so the orbital parity must be even. If the single π has J = 1, so must the pair. But if I-spin is not conserved it turns out (exercise) that you need J = 2 both for the single π and for the pair. Thus I and C violating 3π decays will be quite forbidden.

Consider next the ϕ-meson:

$$\phi \to K^+ + K^-$$

with J = 1 and C = -1. Hence for normal statistics of the K's

$$\phi \not\to K_1 + K_1$$

only

$$\phi \to K_1 + K_2 .$$

Similarly, for the ρ-meson

$$\rho \to 2\pi,$$

again a P state, with definite C. The only meson offering a good possibility of an observation of C-violation is the η. If C is strongly violated, the η can decay strongly. If C is violated then

$$\eta \to 3\pi$$

is possible, even though I-spin may be conserved. Moreover this is completely allowed by angular momentum conservation. It is still an open question whether it exists. The observation would be in a π^+, π^- asymmetry brought about by interference between the C conserving - I violating amplitude and the C violating - I conserving one.

The suggestion that C might be violated in strong interactions came from Prentki and Veltman,[16] inspired presumably by the recent CP violation experiment.[4] They proposed that C-violation in strong interactions would manifest itself in this way in weak interactions, though we would have a hard time observing it in strong interactions directly. The violation was supposed to be medium strong, of the same order of magnitude as the SU_3 breaking. Cabibbo suggested even stronger violations. Bernstein, Feinberg and Lee[17] suggested explictly that the violations might be electromagnetic. The current would be of the form

$$j_\mu = j_\mu^{good} + j_\mu^{bad}$$

where j_μ^{bad} would have the anomalous behaviour

$$C_{op}\, j_\mu^{bad}\, C_{op}^{-1} = + j_\mu^{bad},$$

which as we have seen is very hard to do. You can't do it with two spin - $\frac{1}{2}$ particles. You can't do it easily with minimal interactions (you can, with "minimal" interaction of spin 3/2 particles[17]). If this happens to order α, then in the first place it can account for the $K_2^0 \to 2\pi$ experiment, and in the second place it allows a substantial decay mode

$$\eta \to \pi^0 + e^+ + e^-.$$

All this awaits further experiments.

3.3 TCP-INVARIANCE

We shall consider for a moment the very simplest case namely the scattering of two scalar particles into two others. Each particle is its own charge conjugate. This example will show that relativistic invariance in no way imposes TCP invariance. Because the reaction is its own charge conjugate, C hold automatically. Since the particles are scalars, the scattering amplitude is unable to exhibit any pseudo-scalar behaviour, and P also holds automatically. In fact, the invariant Møller amplitude M may be regarded as a function of the incoming energy W and the angle θ between the incoming and the outgoing particles, say in the CM system. Both of these quantities are scalars. (We could equally well write M in terms of the invariant quantities s and t, and reach the same conclusion). But what about T. If the direct process is

$$a + b \to c + d,$$

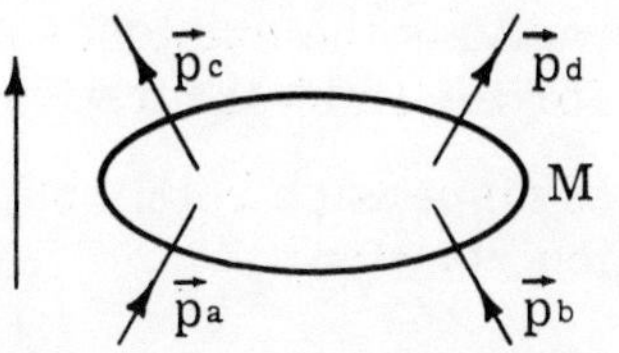

with momenta $\vec{p}_a$, $\vec{p}_b$, $\vec{p}_c$, $\vec{p}_d$, then the time-reversed process is

$$c + d \to a + b$$

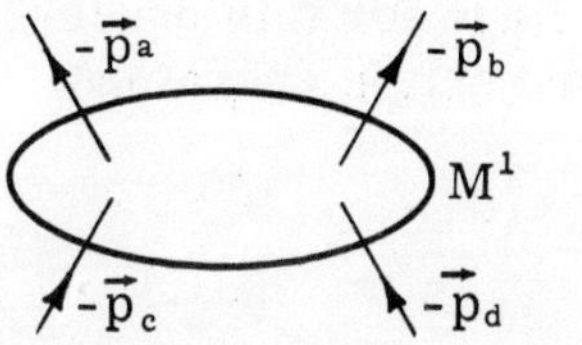

with momenta $-\vec{p}_c$, $-\vec{p}_d$, $-\vec{p}_a$, $-\vec{p}_b$, and we may write the Møller amplitude M' for this process as a function of the same variables W and cos Θ, but no amount of Lorentz invariance can tell you anything about whether M' and M are the same functions of their arguments. They simply can be different functions. Extra assumptions are needed to ensure TCP, and those assumptions might be wrong. These assumptions come naturally out of the field—theoretical formalism or the S-matrix formalism, which has in it the necessary minimum extracted from field theory, with enough added to give the result.

These formalisms are then broad enough to include T-violation, C-violation or P-violation separately, but not TCP-violation. So we hold on desperately to TCP, because if we lost it we should not know what to do. The point is that TCP comes not from Lorentz invariance, but from the general structure of the theory with which we have to deal.

This general structure has had great successes. Among these I should mention

1) The existence of anti-particles. These antiparticles have been discovered, and have the same masses as the particles. This may be regarded as evidence for C-conservation, but if C is violated, it becomes evidence for TCP. In this context electrodynamics should be examined separately, because there is a variety of detailed support for the theory, but in the strong interactions the existence of antiparticles is the main evidence for the field-theoretical formalism.

2) Verification of dispersion relations, especially the pion-nucleon dispersion relations.

Let us consider briefly a more complicated case, say

$$\pi^- + p \to \pi^0 + n.$$

The relativistic amplitude for this process is conventionally written in terms of projection operators, say

$$-A + \frac{i}{2}\gamma(q_1 + q_2)B,$$

and there are rules which tell us what happens to γ-matrices and momenta under PCT, but there is nothing which tells us that the A and the B are the same for this process as for the time-reversed process

$$\pi^0 + n \rightarrow \pi^- + p.$$

So there are two things which have to be put in. One is the algebra of Lorentz transformations, which invites the conclusion that TCP holds. The other is a physical assumption about the invariant coefficients A and B in the amplitude. The S matrix people will put it this way--that A and B are functions of the channel mass s and cross-channel mass (momentum transfer) t and that each is the same function of s and t for all physical values of these variables. This physical assumption can be put much more simply in the field theory—you essentially have to assume space like commutativity for field operators. Our S-matrix proof will be harder, because the S-matrix has an unknown function in it, but it will include the Lagrangian case (in which the unknown function is a local function and there almost is no problem). Of course, if I-spin invariance holds then some processes, for example the two πN processes written down above, become the same process, and some T-invariance is thereby built into the theory, but in general an extra assumption is necessary.

Confining ourselves to spin - $\frac{1}{2}$ fermions,* and arbitrary spin bosons, we may distinguish two cases

(a) number conservation for spin - $\frac{1}{2}$ particles: Q, B, Y, L (lepton number) conserved.

(b) no number conservation for spin - $\frac{1}{2}$ particles (in this case, which is left as an exercise, the arbitrariness in the phase, which we shall find, degenerates to $\neq 1$). It turns out that charge conservation, for example, although perhaps necessary for zero-mass photons, is not necessary for TCP.

*The proof can of course be extended to higher spin fermions, but becomes much more complicated. Since arbitrary spin particles can be formed from spin $\frac{1}{2}$ ones, of course, it is really enough in principle to consider the latter case only.

We shall work out Case (a) for four particle interactions only. The generalization to more particles is obvious. Consider the case of two fermion fields interacting with two boson fields. Then the invariant 2 body → 2 body Møller matrix may be written (with our representation of the γ-matrices).

$$M = \int :\psi_a^\dagger(x_1)\;\beta G(x)\;\psi_b(x_2)\,\phi_c(x_3)\,\phi_d(x_4)\,dx \quad (3.10)$$

where $dx = d^4x_1\; d^4x_2\; d^4x_3\; d^4x_4 \ldots$ and integration is over all values. Here G is any covariant formed from γ- matrices and may depend on all the four-vectors involved in the problem. If one of the ϕ's is a vector or a tensor there has to be a compensating tensor in G or in one of the ϕ's which is contracted with it so that M is a scalar. The notation : : stands for Wick's normal ordering convention.

Before going on to the calculation, let us check exactly what TCP conservation requires of the M-matrix. The connection between the M-matrix and the S-matrix is given by

$$S = 1 + i\,\frac{(2\pi)^4}{\sqrt{\text{product of energies by conventional factors}}}\,M. \quad (3.11)$$

(the usual four dimensional $\delta^4(\Sigma p)$ is contained in M by translational invariance of G. What does time reversal do to S, if T conservation holds? From p. 22 we recall equation (2.27)*:

$$TST^{-1} = S^{-1} = S^\dagger$$

The denominator on the right of (3.11) is real and time-reversible and so goes into itself under T. The numerator also goes into itself, but the coefficient i changes sign. On the other hand, if we take the adjoint of (3.11), the i again changes sign. Therefore

---

*In this section we drop the subscript $_{op}$ from T, C, and P.

$$TST^{-1} = 1 - i(\text{same quotient})\, TMT^{-1}$$

while

$$S^{\dagger} = 1 - i(\text{same quotient})\, M^{\dagger}$$

so that

$$TMT^{-1} = M^{\dagger}. \tag{3.12}$$

If parity and charge conjugation invariance are also to hold, then instead of (3.12) we have to verify

$$(TCP)M(TCP)^{-1} = M^{\dagger}. \tag{3.13}$$

In the case of Lagrangian field theory, $\mathcal{L}$ is self-adjoint and the condition is simply

$$(TCP)\mathcal{L}(TCP)^{-1} = \mathcal{L}.$$

Let us calculate what T, C, and P do to the various fields which come into M. From (2.13), we have, always using our standard representation,

$$P\,\psi_a(\vec{x}, t)\,P^{-1} = \eta_P^a\,\beta\,\psi_a(-\vec{x}, t), \tag{3.14}$$

(here a is a particle label, not a spinor index), and correspondingly

$$P\psi_a^{\dagger}(\vec{x}, t)\; P^{-1} = \eta_P^{a*}\,\beta\,\psi_a^{\dagger}(-\vec{x}, t). \tag{3.14†}$$

For a one-component boson field, we have simply

$$P\phi(\vec{x}, t)P^{-1} = \eta_P\,\phi(-\vec{x}, t); \tag{3.15}$$

for a tensor field, we have to take account both of the imaginary character of components indexed by an odd number of 4's and the different transformation properties of space and time components. I illustrate these points by treating the vector field—the general case then follows easily. It is convenient to introduce an indicator g_μ, defined by

$$g_1 = g_2 = g_3 = -\,g_4 = 1.$$

Then the coordinate transformation of a vector field A_μ may be included in such a formula as (3.15) by writing

$$PA_\mu(\vec{x},t)\,P^{-1} = +\,g_\mu\,\eta_p\,A_\mu(-\vec{x},t)\,. \qquad (3.15^\mu)$$

Let us now apply charge conjugation. From (2.41) and (2.47) we have for the spinor field*

$$(CP)\psi_a(\vec{x},t)\,(CP)^{-1} = \eta_p^a\;\eta_c^a\,\beta\,\rho_2\,\sigma_y\,\psi_a^\dagger(-\vec{x},t) \quad (3.16)$$

and

$$(CP)\,\psi_a^\dagger(\vec{x},t)\,(CP)^{-1} = \eta_p^{a*}\;\eta_c^{a*}\;\beta\rho_2\sigma_y\,\psi_a(-\vec{x},t) \quad (3.16^\dagger)$$

while for the boson fields we have

$$(CP)\,\phi(\vec{x},t)\,(CP)^{-1} = \eta_p\,\eta_c\,\phi^\dagger(-\vec{x},t) \qquad (3.17)$$

and

$$(CP)\,A_\mu(\vec{x},t)\,(CP)^{-1} = g_\mu\eta_c\,g_\mu\eta_p\,A_\mu^\dagger(-\vec{x},t) = \eta_c\,\eta_p\,A_\mu^\dagger(-\vec{x},t) \qquad (3.17^\dagger)$$

(where the second g_μ, which cancels the first, arises from the imaginary sign of the 4th component). Finally, from (2.39),

$$\begin{aligned}
(TCP)\,\psi_a(\vec{x},t)\,(TCP)^{-1} &= -\eta_p^{a*}\eta_c^{a*}\beta\rho_2\eta_T^{a*}\psi_a^\dagger(-\vec{x},-t)\\
(TCP)\,\psi_a^\dagger(\vec{x},t)\,(TCP)^{-1} &= \eta_p^a\,\eta_c^a\,\beta\rho_2\eta_T^a\,\psi_a(-\vec{x},-t) \qquad (3.18)\\
(TCP)\,\phi_d(\vec{x},t)\,(TCP)^{-1} &= \eta_p^{d*}\;\eta_c^{d*}\eta_T^{d*}\phi_d^\dagger(-\vec{x},-t)
\end{aligned}$$

(and the same formula for a vector or tensor boson field, since again A_0 and $\vec{A}$ transform oppositely, whereas T changes i to -i in A_μ). We abbreviate, for each field

---

*There is a hidden assumption here too, to wit, that the transformation (3.16) is unitary, which it will only be if the asymptotic field ψ has the appropriate degrees of freedom corresponding to anti-particle states, and that these enter the S-matrix as given by (3.10).

$$\eta^a_P \eta^a_C \eta^a_T = \delta_a .$$

Substituting into (3.10) and remembering that $\beta\rho_2 = -i\rho_1$, we find quite straightforwardly

$$(TCP)\,M(TCP)^{-1} = \delta_a \delta_b^* \delta_c^* \delta_d^* \int \rho_1 \psi_a(-x_1)\,\beta G^*(x)\,\rho_1 \psi_b^\dagger(-x_2)\,\phi_c^\dagger(-x_3)\,\phi_d^\dagger(-x_4)\,dx$$

$$= -\,\delta_a \delta_b^* \delta_c^* \delta_d^* \int \psi_a(x_1)\,\beta\,[\rho_1 G^*(-x)\rho_1]\,\psi_b^\dagger(x_2)\,\phi_c^\dagger(x_3)\,\phi_d^\dagger)\,x_4)\,dx \tag{3.19}$$

We next calculate $M^\dagger$. In so doing, we have to remember the anticommutation properties of the spinor fields. For this purpose, we remember where necessary that the operator products are normally ordered. From (3.10) we get

$$M^\dagger = \int \psi_b^\dagger(x_2)\,\widetilde{(\beta G^*(x))}\,\psi_a(x_1)\,\phi_d^\dagger(x_4)\,\phi_c^\dagger(x_3)\,dx$$

$$= -\int \psi_a(x_1)\,(\beta G^*(x))\,\psi_b^\dagger(x_2)\,\phi_c^\dagger(x_3)\,\phi_d^\dagger(x_4)\,dx. \tag{3.20}$$

Comparing (3.19) and (3.20), we conclude (remembering that $\rho_1 = -\gamma_5$, and taking the complex conjugate) that for TCP invariance we must have

$$\delta_a^* \delta_b \delta_c \delta_d \gamma_5 G(-x) \gamma_5 = G(x). \tag{3.21}$$

We now have to make the assumption that the covariant $G(x)$ does not depend explicitly on t. For example $G(x)$ could depend on sign (t) (inside the light cone). In S-matrix theory this assumption is usually stated in momentum space: equation (3.21) is

$$\delta_a^* \delta_b \delta_c \delta_d \gamma_5 G(-p) \gamma_5 = G(p)$$

and the assumption is that G is constructed from invariants (see p. 56). (Note that the pseudo-scalar invariant $\epsilon_{\mu\nu\lambda\eta}\, p_{1\mu}\, p_{2\nu}\, p_{3\lambda}\, p_{4\eta}$ which may appear in a $2 \to 3$ body process is invariant under the transformation $p \to -p$.)

To complete the argument, we should discuss the form of G. This function must be built up from covariants, namely (in momentum space)

$$\gamma_5, \gamma_\mu, p_\mu, \gamma \cdot p, p_a \cdot p_b,$$

and other scalars and pseudoscalars constructed from these. G may be as complicated as you please, and nothing is said about its reality--the phase of G has gone out. But, by our explicit assumption, as stated above, $\theta(p_0) = p_0 / |p_0|$ is not permitted.

For every uncontracted vector index—each p_μ or γ_μ which is not dotted into something else— there is a sign change. These sign changes must be compensated. The phases-η's and δ's-are not fixed, and we may choose them to compensate the sign changes. But if TCP is to hold generally, not only for certain particular reactions, we must choose the phases once for all so that the sign changes are compensated in all cases. The choice does not have to be unique.

This choice may be made as follows: to each tensor of rank J among the ϕ's, there will be J uncontracted vector indices in G, therefore a need for J changes of sign. It follows that compensation may be achieved by the choice

$$\delta_c = (-1)^{J_c}$$
$$\delta_d = (-1)^{J_d}$$
$$\delta_a = \delta_b, \qquad |\delta_a| = 1.$$

It is easy to see, and has no doubt been proved, that if there is a particle conservation law, then the M-matrix can be written as a sum of bilinear covariants of the form we have considered. But if particle number is not conserved, in particular, if fermion number is not conserved a further argument is necessary (see exercise, p. 529). I emphasize again that it is easy to write down interactions, such as,

$$\bar{\psi}\psi + \psi C_1 \psi$$

which are Lorentz invariant but not particle-number conserving.

To conclude, a remark about the experimental evidence: should CP hold, except at the level indicated by the Fitch experiment,[9] then we know no more about the validity of TCP than about the validity of CP. If CP should fail above the level of the weak interactions either in the medium strong interactions at the SU_3 —violating level, or in the electromagnetic interactions (cf ref. 23) then we know from the equality of the $K\bar{K}$ masses, that TCP is much better than CP, and the $K_1 K_2$ mass difference puts an upper bound on TCP-violation.

3.4. SU_3

In discussing SU_3 I have two general comments to make at the beginning, one which implies that it is better than it looks, one, that it is worse. It is better than it looks in the sense that the π-meson mass is so small. On the scale of Bev^2, m_π^2 is almost exactly 0.02, while $m_K^2 \sim 0.25$ and $m_N^2 \sim 1$, so that in calculations involving anything heavier, m_π^2 may be neglected. If you think of the mass arising from the difference between some larger mass and a binding energy, then the binding energy of the π-meson is a few hundred Mev greater than the binding energy of the K or the η, which is not very great if you are binding objects of mass a few Bev. So if you are looking at things from the SU_3 point of view, in terms of quarks of one kind or another, the binding energies of the mesons are comparable, and the smallness of the π mass may be regarded as accidental, whereas if you compare the masses, or the masses squared, you get numbers of different orders of magnitude. But as a result, the whole world looks totally asymmetric with respect to SU_3—the π is quite a light particle which mediates a long range force and into which other things can decay. One might say that the remarkable existence of nuclear physics depends on the smallness of the π mass. The binding energies in nuclear physics is of order

$$\epsilon \sim \frac{m_\pi^2}{2m_N} \sim 12 \text{ Mev or so.}$$

Failing the π, the nuclear binding energies would be comparable with the "particle" masses.

The second comment, which I shall expand later, concerns the criticism by Yang and Oakes[18] of the mass formula. Many people jumped on Yang and Oakes—which was great sport, because after all one seldom has the opportunity to jump on Yang—they were technically wrong, but at the same time qualitatively correct. The point was that the location of a state of great width—the N*(1238), for example—can undergo a large shift, if it is just bound rather than just unbound; what is a width in a resonance becomes a shift in a bound state, and so you get large non-perturbative shifts, in the decuplet, for example. The jumpers pointed out that although the width effects have to be taken into account in any real calculation of decuplet levels, it happens that the shifts all cancel out, in a way which as far as anyone can tell today is accidental. I emphasize that this is a non-perturbative effect which involves dynamics in a very intimate, not entirely self-consistent way—one can imagine a pi octet and a nucleon octet whose masses and widths are unitary symmetric (i.e., conform to the mass formula) but which are such that the decuplet shifts do not cancel. This is not something which has been understood in terms of a basic underlying structure; something else is happening which we do not understand, and as long as this is so we cannot expect that the enormous success of SU_3 in the low energy mass spectrum will automatically continue as we go to higher masses.

Algebra of SU_3

Next I shall develop the algebra of SU_3, working in terms of irreducible tensors. This presentation does not pretend to be complete--there are many substantial treatments in the literature[19]--but it includes those features which are essential for the physical applications which follow. The group SU_3 is the group of unitary transformations of three-dimensional complex vectors, with determinant 1.

$$\xi'_{\alpha} = U_{\alpha}^{\beta} \xi_{\beta} \tag{3.22}$$

(summation from 1 to 3 over repeated Greek indices is understood) where U_α^β is a matrix with the properties

$$U_\alpha^\beta U_\beta^{\dagger\gamma} = U_\alpha^\beta U_\gamma^{*\beta} = \delta_\alpha^\gamma$$
$$\det(U_\alpha^\gamma) = 1. \tag{3.22}$$

The complex conjugate matrix gives another (inequivalent!) representation:

$$\eta'^\alpha = U^{\dagger\alpha}_\gamma \eta^\gamma = U_\alpha^{*\gamma}\eta^\gamma \tag{3.24}$$

and because of the unitary property the two representations are contragredient, which means that contraction of an upper and a lower index is invariant (lower indices, which transform like (3.22) are called covariant, upper indices, which transform like (3.24), contravariant). From (3.22) - (3.24), we have

$$\xi'_\alpha \eta'^\alpha = U_\alpha^\beta \xi_\beta U_\gamma^{\dagger\alpha} \eta^\gamma = \delta_\gamma^\beta \xi_\beta \eta^\gamma = \xi_\alpha \eta^\alpha. \tag{3.25}$$

A tensor of covariant rank p and contravariant rank q, or of type (p, q), is a set of quantities $T^{\beta_1\ldots\beta_q}_{\alpha_1\ldots\alpha_p}$ which transforms like ξ_α on each lower index and like η^α on each upper index:

$$T'^{\beta_1\ldots\beta_q}_{\alpha_1\ldots\alpha_p} = U_{\alpha_1}^{\gamma_1} \cdots U_{\alpha_p}^{\gamma_p} U^{\dagger\,\beta_1}_{\delta_1} \cdots U^{\dagger\,\beta_q}_{\delta_q} \quad T^{\delta_1\ldots\delta_q}_{\gamma_1\ldots\gamma_p} \;. \tag{3.26}$$

(The rank of a tensor, which may be anything, should not be confused with the rank of the group, which in the case of SU_3 is 2).

The representation under which an arbitrary tensor transforms is not (in general) irreducible. We shall identify the irreducible tensors in a moment. First we observe that besides the Kronecker delta, δ_α^β, there is here, just as for the rotation group, another numberical tensor which is left unchanged by all transformations, namely the alternating tensor, defined by

$$\epsilon^{\alpha\beta\gamma} = \epsilon_{\alpha\beta\gamma} = \epsilon_{\beta\gamma\alpha} = \epsilon_{\gamma\alpha\beta} = -\epsilon_{\gamma\beta\alpha} = -\epsilon_{\beta\alpha\gamma}, \; \epsilon_{123} = 1, \tag{3.27}$$

which is completely antisymmetric in its indices. The tensor character for these numerical tensors follows from

$$\delta'^{\alpha}_{\beta} = U^{\dagger\alpha}_{\gamma} U^{\delta}_{\beta} \delta^{\gamma}_{\delta} = U^{\gamma}_{\beta} U^{\dagger\alpha}_{\gamma} = \delta^{\alpha}_{\beta},$$

$$\epsilon'_{\alpha\beta\gamma} = U^{\delta}_{\alpha} U^{\epsilon}_{\beta} U^{\zeta}_{\gamma} \epsilon_{\delta\epsilon\zeta} = \epsilon_{\alpha\beta\gamma} \det (U^{\epsilon}_{\delta}) = \epsilon_{\alpha\beta\gamma},$$

$$\epsilon'^{\alpha\beta\gamma} = U^{\dagger\alpha}_{\delta} U^{\dagger\beta}_{\epsilon} U^{\dagger\gamma}_{\zeta} \epsilon^{\delta\epsilon\zeta} = \epsilon^{\alpha\beta\gamma} \det (U^{\dagger\delta}_{\epsilon}) = \epsilon^{\alpha\beta\gamma}.$$

It can be proved[26] that there are no other invariant numerical tensors except for products of these. In dealing with antisymmetric tensors it is often convenient to use an abbreviated notation: If any indices on a tensor expression are surrounded by square brackets [ ], this denotes that with respect to those indices the completely antisymmetric part is to be taken:

$$T_{[\alpha_1 \dots \alpha_p]} = \frac{1}{p!} \sum \pm T_{\alpha_{k_1} \dots \alpha_{k_p}}$$

the sum being taken over all permutations $k_1 \dots k_p$ of $1 \dots p$, and the + (-) sign being attached according as the permutation is even or odd. For example

$$T_{[\alpha\beta]} = 1/2 (T_{\alpha\beta} - T_{\beta\alpha}),$$

and (3.27) may be written

$$\epsilon^{\alpha\beta\gamma} = \epsilon_{\alpha\beta\gamma} = \epsilon_{[\alpha\beta\gamma]}, \quad \epsilon_{123} = 1.$$

A similar notation for the symmetric part, the indices being placed in parentheses, is also useful:

$$T_{(\alpha_1 \dots \alpha_p)} = \frac{1}{p!} \sum T_{\alpha_{k_1} \dots \alpha_{k_p}}$$

the sum being taken over all permutations $k_1 \dots k_p$ of $1 \dots p$, and the + sign being attached to each term.

It is easy to see that antisymmetry is preserved by any (unitary) transformation. It is also seen straightforwardly that the inner product defined by (3.25) makes sense (takes tensors into tensors), even if there are more indices about; from $T_{\alpha_1 \ldots \alpha_p}^{\beta_1 \ldots \beta_q}$ we can form the tensors $T_{\gamma\alpha_2 \ldots \alpha_p}^{\gamma\beta_2 \ldots \beta_q}$, $T_{\alpha_1\gamma\alpha_3 \ldots \alpha_p}^{\gamma\beta_2 \ldots \beta_q}$, $T_{\gamma\alpha_2 \ldots \alpha_p}^{\beta_1\gamma\beta_3 \ldots \beta_q}$, and so on, all of type (p - 1, q - 1). This process of summing over contragredient indices is called contraction or taking a trace.

Indices may be raised and lowered with the help of the alternating tensor. Thus from the covariant vector ξ_α we may form a tensor of contravariant rank 2:

$$\xi^{\alpha\beta} = \epsilon^{\alpha\beta\gamma}\, \xi_\gamma,$$

which is evidently antisymmetric in its indices. This process may be reversed:

$$\xi_\gamma = \frac{1}{2}\, \epsilon_{\gamma\alpha\beta}\, \xi^{\alpha\beta}.$$

Conversely, from a contravariant tensor $\eta^{\alpha\beta}$ we may form a covariant vector

$$\eta_\gamma = \frac{1}{2}\, \epsilon_{\gamma\alpha\beta}\, \eta^{\alpha\beta},$$

but this process is reversible only if $\eta^{\alpha\beta}$ is antisymmetric, because $\epsilon_{\gamma\alpha\beta}$ irreversibly annihilates the symmetric part.

Irreducible tensors

I shall now describe the irreducible tensor representations of SU_3 (it can be shown that every representation is equivalent to a tensor representation). It is clear that a general tensor transformation (3.26) is not irreducible, because a general $T_{\alpha_1 \ldots \alpha_p}^{\beta_1 \ldots \beta_q}$ may be divided into sub-tensors, with particular symmetries of the indices, which transform among themselve. For example, any tensor of contravariant rank 2 may be written

$$T^{\alpha\beta} = \frac{1}{2}\,(T^{\alpha\beta} + T^{\beta\alpha}) + \frac{1}{2}(T^{\alpha\beta} - T^{\beta\alpha}),$$

and the symmetric and antisymmetric parts transform independently. A tensor of type (1, 1) may be written

$$T^{\alpha}_{\beta} = (T^{\alpha}_{\beta} - \frac{1}{3}\,\delta^{\alpha}_{\beta}\,T^{\gamma}_{\gamma}) + \frac{1}{3}\,\delta^{\alpha}_{\beta}\,T^{\gamma}_{\gamma} \tag{3.28}$$

and here the tracefree part and the trace transform independently of one another. This situation is similar to that which arises for the rotation group or for SU_2. The main difference is that in the case of the rotation group upper and lower indices are essentially equivalent, so that we may, for example, consider all indices to be upper, and trace over any pair. Here, upper and lower indices are inequivalent, and we can trace only over contragredient pairs. However, the part of a tensor antisymmetric with respect to any pair of (for example) contravariant indices may be multiplied by $\epsilon_{\alpha\beta\gamma}$ to give another tensor with one more covariant index and two fewer contravariant indices. It turns out that a systematic application of this procedure, together with tracing, yields tensors which transform under irreducible representations: the tensors of all types (p, q), completely symmetric in upper indices together, and in lower indices together, and tracefree on every pair of indices, one upper and one lower, transform irreducibly.

These properties of irreducible tensors may be summed up in the equations

$$\begin{aligned} T^{\beta_1\ldots\beta_q}_{\alpha_1\ldots\alpha_p} &= T^{(\beta_1\ldots\beta_q)}_{(\alpha_1\ldots\alpha_p)} \\ T^{\gamma\beta_2\ldots\beta_q}_{\gamma\alpha_2\ldots\alpha_p} &= 0. \end{aligned} \tag{3.29}$$

The irreducibility of such tensors is intuitively clear; rigorous proofs, which are rather lengthy, may be found in textbooks.[20] They depend on the fact (also provable) that essentially the only possible reduction can be effected by contraction and by symmetry operations on indices, and for tensors with the properties (3.29), neither operation produces anything new.

How many independent components has an irreducible tensor of given type? In enumerating them, we must take into account the symmetry and tracefree conditions (3.29). The symmetry condition is easily dealt with: it implies that the upper indices on a given tensor may be rearranged arbitrarily among themselves without changing the tensor.

The same is true for the lower indices. Therefore a unique component is specified by giving only the <u>numbers</u> of 1's, 2's and 3's among the upper indices say q_1, q_2, q_3, with $q_1 + q_2 + q_3 = q$, and the same information for the lower indices say p_1, p_2, p_3, with $p_1 + p_2 + p_3 = p$. This takes care of the symmetry. Now observe that <u>every</u> tracefree condition may be written in the form

$$T^{3\beta_2\ldots\beta_q}_{3\alpha_2\ldots\alpha_p} = -T^{1\beta_2\ldots\beta_q}_{1\alpha_2\ldots\alpha_p} - T^{2\beta_2\ldots\beta_q}_{2\alpha_2\ldots\alpha_p}$$

and may therefore be regarded as relation which specifies a component with q_3 upper 3's and p_3 lower 3's in terms of components with $q_3 - 1$ upper 3's and $p_3 - 1$ lower 3's. By the repeated use of trace conditions one may therefore express every component with $q_3 > 0$, $p_3 > 0$ in terms of components with $q_3 = 0$ or $p_3 = 0$. On the other hand, all components with either $q_3 = 0$ or $p_3 = 0$ may be specified independently, because none of them may be expressed in terms of others by a tracefree condition (every tracefree condition which does not remove 3's must introduce them). The number of independent components is therefore equal to the number of ways of choosing the p's and q's so that either $p_3 = 0$ or $q_3 = 0$ (or both). If $p_3 = 0$ we have

$$q_1 + q_2 + q_3 = q, \quad p_1 + p_2 = p. \qquad (3.31)$$

We may choose $q_3 = 0, 1, \ldots, q$. Suppose $q_3 = q - q'$ then $q_1 + q_2 = q'$ and we may choose $q_2 = 0, 1, \ldots, q'$, which fixes q_1. The number of ways of choosing q_2 is $q' + 1$, and q' ranges from 0 to q. Hence the number of choices of the n's is

$$\sum_{q'=0}^{q} (q' + 1) = \sum_{q'=1}^{q+1} q' = \frac{1}{2}(q + 1)(q + 2). \qquad (3.32)$$

Similarly the number of choices of the p's is $(p + 1)$. Altogether there are therefore $\frac{1}{2}(q + 1)(p + 1)(q + 2)$ independent components with $p_3 = 0$. Similarly, interchanging the roles of p and q, there are $\frac{1}{2}(q + 1)(p + 1)(p + 2)$ independent components with $q_3 = 0$. The components with $p_3 = q_3 = 0$ have been counted twice. There are $(p + 1)(q + 1)$ of them. The total number of independent components of an irreducible tensor of type (p, q) is therefore

$$n_{p,q} = \frac{1}{2}(p+1)(q+1)(q+2) + \frac{1}{2}(q+1)(p+1)(p+2) - (p+1)(q+1)$$
$$= \frac{1}{2}(p+1)(q+1)(p+q+2). \tag{3.33}$$

There is a conventional abbreviation for an irreducible representation by tensors of given type. If $p \geq q$, the representation is denoted $n_{p,q}$; if $q > p$, it is denoted $\bar{n}_{q,p}$. The table shows the notation for some of the irreducible representations of low dimension (sometimes one puts the n in curly brackets $\{1\}$, $\{3\}$, $\{\bar{3}\}$, etc.)

Type	Notation
(0, 0)	1
(1, 0)	3
(0, 1)	$\bar{3}$
(1, 1)	8
(3, 0)	10
(0, 3)	$\overline{10}$
(2, 2)	27
(4, 1)	35
(1, 4)	$\overline{35}$
etc.	

Reduction of the Direct Product

In the applications we have often to multiply two tensors together. However, the product of irreducible tensors is not (unless one of them is a scalar) irreducible. For example, the product of two vectors of contravariant rank 1 is

$$S^\alpha T^\beta = \frac{1}{2}(S^\alpha T^\beta + T^\alpha S^\beta) + \frac{1}{2}(S^\alpha T^\beta - T^\alpha S^\beta)$$
$$= S^{(\alpha}T^{\beta)} + \frac{1}{2}\epsilon^{\alpha\beta\gamma}U_\gamma$$

where $U_\gamma = \epsilon_{\gamma\delta\epsilon} S^\delta T^\epsilon$ is a vector of covariant rank 1. Thus the product may be reduced to an irreducible tensor of type

(0, 2) and one of type (1, 0). Symbolically this is written

$$(0,\ 1) \oplus (0,\ 1) = (0,\ 2) \oplus (1,\ 0),$$

or in terms of the abbreviated violation,

$$\bar{3} \oplus \bar{3} = \bar{6} \oplus 3$$

This process--writing the product of irreducible representations as a sum of irreducible representations— is called the reduction of the direct product. The coefficients of the sum are called Clebsch-Gordan coefficients, and the sum itself is called the Clebsch-Gordan series. The subject has a large literature,[19] and there are several methods for carrying out the reduction in the general case. Two of them will be described here. The method of Young diagrams, which has index applications,[21] will be explained here, in the special form in which it is applicable to SU_3.[22] Another simple method will be described later on (p. 95).

A Young diagram, or tableau, is an array of symbols in horizontal and vertical rows of varying length, which corresponds to a tensor of given symmetry. Consider a tensor with upper indices, and let $(q, \dots q_r)$ be a partition of q, that is, a set of positive integers, whose sum is $q : q_1 + q_2 + \cdots + q_r = q$ arranged in decreasing order: $q_1 \geq q_2 \geq \dots \geq q_r$. The corresponding diagram has q_k spaces in the kth row, $k = 1, 2, \dots, r$, the first space of each row being directly under the first space of the preceding row. If the available spaces are filled with the q indices of the tensor, the connection between diagram and tensor symmetry is prescribed by the

RULE: symmetrize over the indices in each row, and then antisymmetrize over the indices in each column.

For example, for tensors with two upper indices, there are just two diagrams,

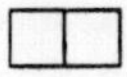 and 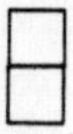

The indices α, β may be inserted in either order in the two diagrams, and the corresponding tensors are just the symmetric and antisymmetric tensors of contravariant rank 2. A less trivial example is the case of three upper indices. There are three possible diagrams:

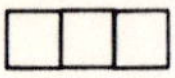 , 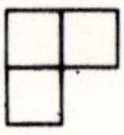 and 

(The diagram  is not allowed, since it does not have $q_1 \geq q_2$). Again the order of the indices in each row and in each column is immaterial, but the division between the rows (or columns) does matter in the case of the middle diagram, where two essentially different arrangements of indices are 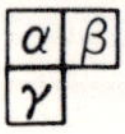 and 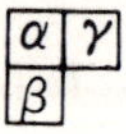

If $T^{\alpha\beta\gamma}$ is an arbitrary tensor, the tensors which may be formed from it with the three diagrams are:

$$T_1^{\alpha\beta\gamma} = T^{(\alpha\beta\gamma)}$$

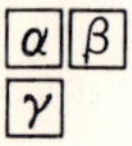

$$T_2^{\alpha\beta\gamma} = T^{\alpha\beta\gamma} + T^{\beta\alpha\gamma} - T^{\gamma\beta\alpha} - T^{\beta\gamma\alpha}$$

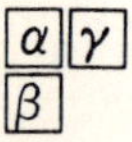

$$T_3^{\alpha\beta\gamma} = T^{\alpha\beta\gamma} + T^{\gamma\beta\alpha} - T^{\beta\alpha\gamma} - T^{\gamma\alpha\beta}$$

$$T_4^{\alpha\beta\gamma} = T^{[\alpha\beta\gamma]}.$$

It is shown in the literature[23] that each of these tensors transforms irreducibly under the group of general (nonsingular) linear transformations (the "general linear" group GL(n)) but under a subgroup, the irreducibility may not always survive, and in the case of SU_3 it does not. The reducibility under SU_3 of all types below T, arises from the possibility of converting upper to lower indices with $\epsilon^{\alpha\beta\gamma}$.

For example, if $T'^{\gamma}_{\delta} = \frac{1}{2}\epsilon_{\delta\alpha\beta} T_3^{\alpha\beta\gamma}$, then

$$T_3^{\alpha\beta\gamma} = \epsilon^{\alpha\beta\delta} T'^{\gamma}_{\delta}$$

while if

$$T' = \frac{1}{6}\epsilon_{\alpha\beta\gamma} T_4^{\alpha\beta\gamma}, \text{ then}$$

$$T_4^{\alpha\beta\gamma} = \epsilon^{\alpha\beta\gamma} T'.$$

Of course the symmetries of T_2 and T_3 are related. However both are needed if $T^{\alpha\beta\gamma}$ is to be expressed in terms of the $T_i^{\alpha\beta\gamma}$ but the choice made above is not unique. Such an expression is

$$T^{\alpha\beta\gamma} = T_1^{\alpha\beta\gamma} + \frac{1}{3} T_2^{\alpha\beta\gamma} + \frac{1}{3} T_3^{\alpha\beta\gamma} + T_4^{\alpha\beta\gamma}.$$

The use of Young diagrams depends in general on the fact that an arbitrary tensor with any number q of upper (lower) indices may be written in terms of tensors constructed from diagrams with q spaces.

Evidently the problem of reduction of the direct product may be expressed in terms of the corresponding diagrams; before stating the solution we deal with some special features which arise in the case of SU_3. An arbitrary irreducible tensor $T^{\beta_1\ldots\beta_q}_{\alpha_1\ldots\alpha_p}$ may be replaced by a tensor with lower (or upper) indices only, by multiplying by ϵ' s. For example, let

$$T'_{\alpha_1\ldots\alpha_p\alpha_{p+1}\ldots\alpha_q\gamma_1\ldots\gamma_q} = \epsilon_{\alpha_{p+1}\gamma_1\beta_1} \cdots \epsilon_{\alpha_{p+q}\gamma_q\beta_q} T^{\beta_1\ldots\beta_q}_{\alpha_1\ldots\alpha_p}$$

Then T' must be irreducible under SU_3, since T is irreducible. A fortiori, T' must be irreducible under the general linear group. What is its Young diagram? It is easily seen to be

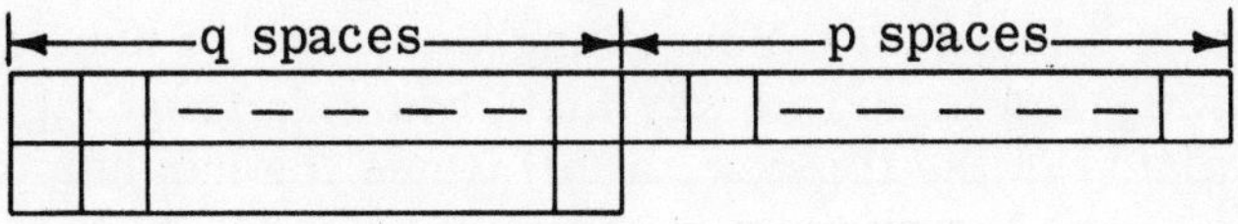

which for T' as written above may be filled in with

$$\begin{matrix} \alpha_{p+1} \ldots \alpha_{p+q} & \alpha_1 \ldots \alpha_p \\ \gamma_1 \ldots \gamma_q & \end{matrix}$$

Another special feature of SU_3, already noticed in the example above, is that in a Young diagram any column with three rows defines a scalar (and of course any column with more than three rows gives zero, since for a non-zero contribution under antisymmetrization, indices must take distinct values, while the indices involved in this case take only the three values 1, 2, 3.)

With these special features in mind, we may adapt to SU_3 the rules for reduction of a direct product under GL(n). The adapted rules may be stated as follows:

To construct $(p, q) \otimes (p', q')$,

1) Write down the diagram for (p, q), denoting the symbols, say, by dots:

$$\begin{array}{ll} \longleftarrow q \longrightarrow & \longleftarrow p \longrightarrow \\ \bullet\ \bullet\ \cdots\ \bullet & \bullet\ \bullet\ \cdots\ \bullet \\ \bullet\ \bullet\ \cdots\ \bullet & \end{array}$$

Write down (or imagine) also the diagram for (p', q'), denoting the symbols in the first row by α's and the symbols in the second row by β's:

$$\begin{array}{ll} \longleftarrow q' \longrightarrow & \longleftarrow p' \longrightarrow \\ \alpha\,\alpha\ \cdots\ \alpha & \alpha\,\alpha\ \cdots\ \alpha \\ \beta\,\beta\ \cdots\ \beta & \end{array}$$

2) Add the $(p' + q')$ α's to the other diagram in such a way as to form a Young diagram with at most three rows, but with no two α's under each other.

3) Add the q' β's to the other diagram so that the augmented diagram is a legitimate Young diagram (rows of non-increasing length), and in such a way that, counting from right to left, starting in the top row, at all times the number of α's is no less than the number of β's.

4) Drop all columns of three, write the tensor corresponding to the remaining diagram, symmetrize the row indices and antisymmetrize the column indices. Raise the anti-symmetrized pairs in the ϵ's.

5) Repeat this procedure, adding the α's and β's in all possible ways consistent with the rules stated. The resulting tensors constitute the required reduction.

An example will show how the rules are applied in simple cases:

Example 1. (1, 1) $\otimes$ (1, 1) or 8 $\otimes$ 8.

```
Rule 1)     . .              α        α
            .                β

Rule 2)     . . α α      . . α    . . α    . .
  +5)       . .          . α      .        . α
                                  α        α

Rule 3)     . . α α      . . α    . . α    . .
  +5)       . β          . α β    . β      . α
                                  α        α β

            . . α α      . . α
            .            . α
            β            β

Rule 4)     .  .  α  α
  +5)       .  β             ⇒        (2, 2) = 27

            .  .  α
            .  α  β          ⇒        (0, 3) = 1̄0̄

            .  .  α         .  α
            .  β      ⇒     β  ⇒      (1, 1) = 8
            α

            .  .
            .  α      ⇒      no diagram ⇒ (0, 0) = 1
            α  β

            .  .  αα        .  αα
            .         ⇒               ⇒   (3, 0) = 10
            β
```

(In the second Rule 4 line, the result is $(0, 3) = \overline{10}$.)

$$\begin{matrix} \cdot & \cdot & \alpha \\ \cdot & \alpha & \\ \beta & & \end{matrix} \Longrightarrow \begin{matrix} \cdot & \alpha \\ \alpha & \end{matrix} \Longrightarrow (1,\ 1) = 8$$

from which we get the familiar rule

$$8 \otimes 8 = 27 \oplus 10 \oplus \overline{10} \oplus 8 \oplus 8 \oplus 1. \tag{3.34}$$

Generators of SU_3

In order to develop any physical interpretation of SU_3, we must know the group algebra, which is to say that we need the generators of the group. To obtain the group algebra, we consider a transformation near the identity, say

$$U_\alpha^\beta = \delta_\alpha^\beta + i\,\epsilon\, u_\alpha^\beta, \tag{3.35}$$

where ϵ is a small parameter whose square and higher powers may be neglected.

The Hermitian adjoint of U_α^β is $U_\alpha^{\dagger\beta}$, given by

$$U_\alpha^{\dagger\beta} = \delta_\alpha^\beta - i\,\epsilon\, u_\beta^{*\alpha}.$$

and so the unitary condition (3.23) is

$$\delta_\alpha^\gamma = U_\alpha^\beta\, U_\beta^{\dagger\gamma} = (\delta_\alpha^\beta + i\,\epsilon\, u_\alpha^\beta)(\delta_\beta^\gamma - i\epsilon u_\gamma^{*\beta})$$

which gives

$$u_\alpha^\beta - u_\beta^{*\alpha} = 0, \tag{3.36}$$

so that the matrix $u = \{u_\alpha^\beta\}$ must be Hermitian
The unimodular condition $\det(U_\alpha^\gamma) = 1$ is, to first order,

$$1 + i\epsilon\, u_\alpha^\alpha = 1$$

so that

$$u_\alpha^\alpha = 0, \tag{3.37}$$

and the Hermitian matrix u must therefore also be tracefree. This matrix has n^2 complex elements, hence depends on $2n^2$

real numbers. Equation (3.36) says that the real part of u is a symmetric matrix and the imaginary part an anti-symmetric matrix. This reduces the number of independent real numbers to n^2. The condition (3.37) reduces this by one more, so that finally the general infinitesimal transformation depends on $n^2 - 1$ (= 8 for n = 3) real parameters. Since u is **Hermitian and tracefree, it is often possible to choose these parameters so that u is a linear combination of** Hermitian tracefree matrices.

For SU_3, one standard choice of basis is the set of eight Hermitian tracefree matrices T_1, ..., T_8 defined as follows (the dots denote zeros):

$$
\left.
\begin{array}{lll}
T_1 = \frac{1}{2}\begin{bmatrix} \cdot & 1 & \cdot \\ 1 & \cdot & \cdot \\ \cdot & \cdot & \cdot \end{bmatrix} &
T_4 = \frac{1}{2}\begin{bmatrix} \cdot & \cdot & 1 \\ \cdot & \cdot & \cdot \\ 1 & \cdot & \cdot \end{bmatrix} &
T_6 = \frac{1}{2}\begin{bmatrix} \cdot & \cdot & \cdot \\ \cdot & \cdot & 1 \\ \cdot & 1 & \cdot \end{bmatrix} \\
T_2 = \frac{1}{2}\begin{bmatrix} \cdot & -i & \cdot \\ i & \cdot & \cdot \\ \cdot & \cdot & \cdot \end{bmatrix} &
T_5 = \frac{1}{2}\begin{bmatrix} \cdot & \cdot & -i \\ \cdot & \cdot & \cdot \\ i & \cdot & \cdot \end{bmatrix} &
T_7 = \frac{1}{2}\begin{bmatrix} \cdot & \cdot & \cdot \\ \cdot & \cdot & -i \\ \cdot & i & \cdot \end{bmatrix} \\
T_3 = \frac{1}{2}\begin{bmatrix} 1 & \cdot & \cdot \\ \cdot & -1 & \cdot \\ \cdot & \cdot & \cdot \end{bmatrix} &
 &
T_8 = \frac{1}{2}\begin{bmatrix} \frac{2}{3} & \cdot & \cdot \\ \cdot & \frac{2}{3} & \cdot \\ \cdot & \cdot & -\frac{4}{3} \end{bmatrix}
\end{array}
\right\}
\tag{3.38}
$$

It can be shown that the maximum number of mutually commuting generators for SU_n is n - 1. This number is called the <u>rank</u> of the group. In the basis chosen above, T_3 and T_8 are diagonal, which shows that they are the mutually commuting pair. Also T_1, T_2 and T_3 are set of Pauli matrices on the first two rows and columns, while T_4, T_5 and $\frac{1}{2}(T_3 + 3/2T_8)$ are a set on the first and last row and first and last column, and T_6, T_7 and $\frac{1}{2}(-T_3 + 3/2T_8)$ are a set on the last two rows and columns. Evidently this choice of basis picks out the 3-axis for distinguished treatment. This has no fundamental significance—it is merely a matter of convention.

Let us determine the behaviour of any irreducible tensor under an infinitesimal transformation. From (3.26) we have

$$\begin{aligned} T'^{\beta_1\ldots\beta_q}_{\alpha_1\ldots\alpha_p} &= U^{\gamma_1}_{\alpha_1} \ldots U^{\gamma_p}_{\alpha_p} U^{\dagger\,\beta_1}_{\delta_1} \ldots U^{\dagger\,\beta_q}_{\delta_q} T^{\delta_1\ldots\delta_q}_{\gamma_1\ldots\gamma_p} \\ &= (\delta^{\gamma_1}_{\alpha_1} + i\epsilon u^{\gamma_1}_{\alpha_1}) \ldots (\delta^{\gamma_p}_{\alpha_p} + i\epsilon u^{\gamma_p}_{\alpha_p})(\delta^{\beta_1}_{\delta_1} - i\epsilon u^{\beta_1}_{\delta_1}) \\ &\quad \ldots (\delta^{\beta_q}_{\delta_q} - i\epsilon u^{\beta_q}_{\delta_q}) T^{\delta_1\ldots\delta_q}_{\gamma_1\ldots\gamma_p} = T^{\beta_1\ldots\beta_q}_{\alpha_1\ldots\alpha_p} \\ &\quad + i\epsilon [\, p u^{\gamma}_{(\alpha_1} T^{\beta_1\ldots\beta_q}_{\alpha_2\ldots\alpha_p)\gamma} - q u^{(\beta_1}_{\delta} T^{\beta_2\ldots\beta_q)\delta}_{\alpha_1\ldots\alpha_p} \,] \end{aligned} \tag{3.39}$$

In particular, suppose that u^{γ}_{α} is diagonal:

$$u^{\gamma}_{\alpha} = \sum_k n_k \delta^k_{\alpha} \delta^{\gamma}_k \iff (u^{\gamma}_{\alpha}) = \begin{bmatrix} n_1 & \cdot & \cdot \\ \cdot & n_2 & \cdot \\ \cdot & \cdot & n_3 \end{bmatrix} . \tag{3.40}$$

In order to evaluate the effect of this transformation, recall that an irreducible tensor is completely symmetric in each set of indices, so that the order in which they are written does not matter. Therefore we may identify a particular component by giving the numbers of 1's, 2's and 3's among the indices: let

$$T\begin{pmatrix} q_1 & q_2 & q_3 \\ p_1 & p_2 & p_3 \end{pmatrix}$$

denote that component which has respectively p_1 1's, p_2 2's, p_3 3's among its lower indices ($p_1 + p_2 + p_3 = p$) and q_1 1's, q_2 2's, q_3 3's among its upper indices ($q_1 + q_2 + q_3 = q$). (The different components are not in general all independent, because of the tracefree condition on irreducible tensors, but for the moment that does not matter). Inserting (3.40) into (3.39), we find, with this notation,

$$T'\begin{pmatrix} q_1 & q_2 & q_3 \\ p_1 & p_2 & p_3 \end{pmatrix} = T\begin{pmatrix} q_1 & q_2 & q_3 \\ p_1 & p_2 & p_3 \end{pmatrix} \left[1 + i\epsilon \sum_k n_k (p_k - q_k)\right] . \tag{3.41}$$

Now suppose that $T^{\beta_1 \dots \beta_q}_{\alpha_1 \dots \alpha_p}$ is an operator on a Hilbert space of states. If U is the unitary operator on the Hilbert space which corresponds to the SU_3 transformation U^γ_α, then $T^{\beta_1 \dots \beta_q}_{\alpha_1 \dots \alpha_p}$ is called a <u>tensor operator</u> if

$$UT^{\beta_1 \dots \beta_q}_{\alpha_1 \dots \alpha_p} U^{-1} = U^{\gamma_1}_{\alpha_1} \dots U^{\gamma_p}_{\alpha_p} U^{\dagger\,\beta_1}_{\delta_1} \dots U^{\dagger\,\beta_q}_{\delta_q} T^{\delta_1 \dots \delta_q}_{\gamma_1 \dots \gamma_p}, \tag{3.42}$$

in other words, if the effect of transforming to UTU^{-1} is the same as the effect of the corresponding SU_3 transformation. In particular, if $U^\beta_\alpha = \delta^\beta_\alpha + i\epsilon u^\beta_\alpha$, we may write

$$U = 1 + i\epsilon\,\mathcal{U}$$

where $\mathcal{U}$, which is a Hermitian operator, is the generator of the transformation on Hilbert space corresponding to the SU_3 transformation generated by u^γ_α. Since

$$\begin{aligned} UT^{\vdots\vdots\vdots}U^{-1} &= (\mathbb{1} + i\epsilon\mathcal{U})T^{\vdots\vdots\vdots}(\mathbb{1} - i\epsilon\mathcal{U}) \\ &= T + i\epsilon[\mathcal{U}, T], \end{aligned}$$

equation (3.39) becomes, for a tensor operator,

$$[\mathcal{U}, T^{\beta_1 \dots \beta_q}_{\alpha_1 \dots \alpha_p}] = pu^{\gamma}_{(\alpha_1} T^{\beta_1 \dots \beta_q}_{\alpha_2 \dots \alpha_p)\gamma} - qu^{(\beta_1}_{\delta} T^{\beta_2 \dots \beta_q)\delta}_{\alpha_1 \dots \alpha_p} \tag{3.43}$$

and equation (3.41) may be rewritten

$$[\mathcal{U}, T\begin{pmatrix} q_1 q_2 q_3 \\ p_1 p_2 p_3 \end{pmatrix}] = \left[\sum_k n_k(p_k - q_k)\right] T\begin{pmatrix} q_1 q_2 q_3 \\ p_1 p_2 p_3 \end{pmatrix} \tag{3.44}$$

If $|0\rangle$ denote the vacuum state, assumed invariant under SU_3, so that $U|0\rangle = |0\rangle$ for every unitary operator U corresponding to an SU_3 transformation, then operating on $|0\rangle$ with (3.44) gives

$$U\,T\begin{pmatrix} q_1 q_2 q_3 \\ p_1 p_2 p_3 \end{pmatrix}|0\rangle = \left[\sum_k n_k(p_k - q_k)\right] T\begin{pmatrix} q_1 q_2 q_3 \\ p_1 p_2 p_3 \end{pmatrix}|0\rangle. \tag{3.45}$$

This equation shows that each component of an irreducible tensor operator, acting on the vacuum state, yields a simultaneous eigenstate of the operators corresponding to diagonal generators. The eigenvalues may be read off from (3.45).

We distinguish between the generators of SU_3 transformstions, say the matrices of equation (3.38), and the generators of the corresponding transformations in Hilbert space, by writing the former with their tensor indices, $T_{1\alpha}^{\ \beta}$, ... and the latter simply T_1, ... Where confusion may arise, the Hilbert space generators are written with the subscript $_{op}$ to emphasize their nature. Thus an arbitrary SU_3 transformation has generator

$$U_\alpha^\beta = \sum_{i=1}^{8} a^i T_{i\alpha}^{\ \beta} ,$$

where a^i are any numbers, and the corresponding Hilbert space generator is

$$\mathcal{U} = \sum_{i=1}^{8} a^i T_{i_{op}} = \sum_{i=1}^{8} a^i T_i .$$

Physical Interpretation

We come, at last, to the physical interpretation. We want SU_3 to include isotopic spin, so we begin by identifying T_1, T_2, T_3 as the isotopic spin operators:

$$\vec{I} = (T_1,\ T_2,\ T_3).$$

Since hypercharge should commute with isotopic spin, we may conjecture that

$$Y = T_8, \tag{3.46}$$

or something like it. From (3.38) and (3.45), the eigenvalues of T_8 may be written

$$\frac{1}{3}(p_1 - q_1) + \frac{1}{3}(p_2 - q_2) - \frac{2}{3}(p_3 - q_3) = p_1 - q_1 + p_2 - q_2 - \frac{2}{3}(p - q) \tag{3.47}$$

so that if we take

$$Y = T_8 + (\text{const}).\ \mathbb{1} \tag{3.48}$$

then hypercharges may differ by one unit, as they should. It follows from (3.47) that unless

$$p - q \equiv 0 (\text{mod } 3) \tag{3.49}$$

tensor operators will correspond (via (3.45)) to states of fractional as well as integral hypercharge. In the quark model, this is overcome by adopting (3.46), but identifying with known particles only those states for which (3.49), holds, while in the Sakata model, it is overcome by adopting (3.48) in the form

$$Y = T_8 + \frac{2}{3} \mathbb{1} \tag{3.50}$$

which gives for covariant vectors ($p = 1$, $q = 0$) eigenvalues of Y equal to $p_1 + p_2$.

The same difficulty arises with respect to charge. If we wish to preserve the Gell-Mann-Nisijima formula

$$Q = I_3 + \frac{1}{2} Y$$

then (3.46) gives

$$[Q_\alpha^\beta] = \begin{bmatrix} \frac{2}{3} & . & . \\ . & -\frac{1}{3} & . \\ . & . & -\frac{1}{3} \end{bmatrix}$$

which gives rise to eigenstates of fractional charge except when (3.49) is satisfied. From (3.50), on the other hand, one finds

$$[Q_\alpha^\beta] = \begin{bmatrix} 1 & . & . \\ . & 0 & . \\ . & . & 0 \end{bmatrix}$$

which gives integral charges only. The quantum numbers for the "basic triplet" of 'particle' creation operators (vector operator of type (1, 0)) are therefore (noting that

$$\left(\vec{I}^{2\,\gamma}_{\alpha}\right) = \frac{3}{4}\left(\begin{bmatrix} 1 & \cdot & \cdot \\ \cdot & 1 & \cdot \\ \cdot & \cdot & \cdot \end{bmatrix}\right)$$

Quark Model

Component	$Y = T_8$	$Q = I_3 + \frac{1}{2}Y$	$I(\vec{I}^2 = I(I+1))$
ξ_1	$\frac{1}{3}$	$\frac{2}{3}$	$\frac{1}{2}$
ξ_2	$\frac{1}{3}$	$-\frac{1}{3}$	$\frac{1}{2}$
ξ_3	$-\frac{2}{3}$	$-\frac{1}{3}$	0

Sakata Model

Component	$Y = T_8 + \frac{2}{3}\mathbb{1}$	$Q = I_3 + \frac{1}{2}Y$	I	
ξ_1	1	1	$\frac{1}{2}$	p
ξ_2	1	0	$\frac{1}{2}$	n
ξ_3	0	0	0	λ

The identifications on the right lead to contradiction with experiment if we understand these particles to be the ordinary proton, neutron and Λ. It is however possible to regard them as (stable or unstable) higher mass states.

Gell-Mann - Ne'eman model. Let us see how to construct the baryon octet in the 'eightfold way' of Gell-Mann and Ne'eman. We make the definite identifications

$$\left.\begin{aligned} &\vec{I} = (T_1, T_2, T_3) \\ &Y = T_8 \\ &Q = T_3 + \frac{1}{2}Y = T_3 + \frac{1}{2}T_8 . \end{aligned}\right\} \quad (3.51)$$

From a tracefree tensor operator M_α^β we can with these identifications construct simultaneous eigenstate of I_3 and Y, with eigenvalues given by (3.45), which take the form

$$Y = \delta^1_\alpha + \delta^2_\alpha - \delta_1^\beta - \delta_2^\beta$$

$$I_3 = \frac{1}{2}(\delta^1_\alpha - \delta^2_\alpha - \delta_1^\beta + \delta_2^\beta) \qquad (3.52)$$

and

$$Q = \delta^1_\alpha - \delta_1^\beta$$

A straightforward calculation from (3.43) gives a formula for $\vec{I}^2$ which leads to the values of I given in the table, together with results from (3.52):

Octet States

	M_α^β	Y	I_3	I	Q	Baryon state
	M_1^3	1	$\frac{1}{2}$	$\frac{1}{2}$	1	p
	M_2^3	1	$-\frac{1}{2}$	$\frac{1}{2}$	0	n
	M_3^1	-1	$-\frac{1}{2}$	$\frac{1}{2}$	-1	Ξ^-
	M_3^2	-1	$\frac{1}{2}$	$\frac{1}{2}$	0	Ξ^0
	M_1^2	0	1	1	1	Σ^+
	M_2^1	0	-1	1	-1	Σ^-
only two independent states	M_3^3	0	0	0	0	multiple of Λ^0
	M_2^2	0	0	mixed	0	combinations of Σ^0 & Λ^0
	M_1^1	0	0		0	

The eistenstates of isotopic spin in the last two rows are given by $M_1^1 + M_2^2$, with $I = 0$, and $M_1^1 - M_2^2$, with $I = 1$.

Exercise: Calculate Y and I_3 for the decuplet $(3, 0) = \{10\}$.

Solution: From (3.45),

$$Y = \frac{1}{3}(p_1 + p_2 - 2p_3) = 1 - p_3$$

$$I_3 = \frac{1}{2}(p_1 - p_2).$$ Hence we may calculate

the table:

Decuplet States (Cont'd.)

$T_{\alpha\beta\gamma}$	Y	I_3	Q
T_{111}	1	$\frac{3}{2}$	2
T_{112}	1	$\frac{1}{2}$	1
T_{113}	0	1	1
T_{122}	1	$-\frac{1}{2}$	0
T_{123}	0	0	0
T_{133}	-1	$\frac{1}{2}$	0
T_{222}	1	$-\frac{3}{2}$	-1
T_{223}	0	-1	-1
T_{233}	-1	$-\frac{1}{2}$	-1
T_{333}	-2	0	-1

Returning to the octet case, we deal with the ambiguities connected with M_1^1, M_2^2 and M_3^3 by constructing a scalar product. Notice that in this case the transformation equation(3.42) is

$$UT_\alpha^\beta U^{-1} = U_\alpha^\gamma\, U_\delta^{\dagger\beta}\, T_\gamma^\delta$$

and the Hermitian conjugate of this equation is

$$UT^{\dagger\beta}_{\alpha} U^{-1} = U_\alpha^{\gamma *}\widetilde{U}_\delta{}^\beta\, T_\gamma^\delta = U_\gamma^{\alpha\dagger}\, U_\beta^\delta\, T_\gamma^{\delta\dagger}$$

so that the adjoint operator transforms, on each index, in the opposite way to the original operator.

It follows that a scalar product $(M_\beta^\alpha, M_\delta^\gamma)$ must be a tensor contragredient on β, γ and cogredient on α, δ. Moreover the states with different values of the indices are pure states, and may be normalizes to unity, for example $(M_3^1, M_3^1) = 1$, and are orthogonal to all other states. It can be shown with some work that the only possible scalar product with these properties which conforms also to the tracefree condition $M_\alpha^\alpha = 0$ is

$$(M_\beta^\alpha, M_\delta^\gamma) = \delta_\alpha^\gamma\, \delta_\delta^\beta - \frac{1}{3}\, \delta_\alpha^\beta \delta_\delta^\gamma. \qquad (3.51)$$

With this normalization, all the off-diagonal components are normalized to unity, but for the diagonal states

$$(M_1^1, M_1^1) = (M_2^2, M_2^2) = (M_3^3, M_3^3) = \frac{2}{3}$$

$$(M_1^1, M_2^2) = (M_2^2, M_3^3) = (M_3^3, M_1^1) = -\frac{1}{3}. \tag{3.52}$$

Since $|M_3>$ is an eigenstate of $\vec{I}^2$, with eigenvalue 0, it was identified in the table as a multiple of $|\Lambda>$. It follows from (3.52) that (with a conventional sign) we get the correct normalization if we take

$$M_3^3 = -\sqrt{\frac{2}{3}}\ \Lambda^0 \tag{3.53}$$

If we now assume

$$M_1^1 = \alpha_1 \Sigma^0 + \beta_1 \Lambda^0$$

$$M_2^2 = \alpha_2 \Sigma^0 + \beta_2 \Lambda^0$$

then the trace-free condition gives

$$\alpha_1 + \alpha_2 = 0$$

$$\beta_1 + \beta_2 = \sqrt{\frac{2}{3}}$$

and by taking scalar products we find

$$-\frac{1}{3} = (M_3^3, M_1^1) = -\sqrt{\frac{2}{3}}\ \beta_1 \Rightarrow \beta_1 = \frac{1}{\sqrt{6}} \Rightarrow \beta_2 = \frac{1}{\sqrt{6}},$$

and

$$\frac{2}{3} = (M_1^1, M_1^1) = \alpha_1 \alpha_1^* + \frac{1}{6} \Rightarrow |\alpha_1|^2 = \frac{1}{2} \Rightarrow \alpha_1 = \frac{1}{\sqrt{2}}$$

(with a conventional choice of phase). We now have

$$\left.\begin{aligned} M_1^1 &= \frac{1}{\sqrt{2}}\ \Sigma^0 + \frac{1}{\sqrt{6}}\ \Lambda^0 \\ M_2^2 &= -\frac{1}{\sqrt{2}}\ \Sigma^0 + \frac{1}{\sqrt{6}}\ \Lambda^0. \end{aligned}\right\} \tag{3.54}$$

We now have all the components of M^{α}_{β} identified, and may write this tensor operator in the form of a matrix:

$$B = \begin{bmatrix} \frac{1}{\sqrt{2}}\Sigma^0 + \frac{1}{\sqrt{6}}\Lambda^0 & \Sigma^+ & p \\ \Sigma^- & -\frac{1}{\sqrt{2}}\Sigma^0 + \frac{1}{\sqrt{6}}\Lambda^0 & n \\ \Xi^- & \Xi^0 & -\sqrt{\frac{2}{3}}\Lambda^0 \end{bmatrix} \quad (3.55)$$

Precisely similar arguments lead to the pseudoscalar meson octet and the vector meson octet, which may be written

$$P = \begin{bmatrix} \frac{1}{\sqrt{2}}\pi^0 + \frac{1}{\sqrt{6}}\eta & \pi^+ & K^+ \\ \pi^- & -\frac{1}{\sqrt{2}}\pi^0 + \frac{1}{\sqrt{6}}\eta & K^0 \\ K^- & K^0 & -\sqrt{\frac{2}{3}}\eta \end{bmatrix}$$

$$V = \begin{bmatrix} \frac{1}{\sqrt{2}}\rho^0 + \frac{1}{\sqrt{6}}\phi & \rho^+ & K^{*+}(891) \\ \rho^- & -\frac{1}{\sqrt{2}}\rho^0 + \frac{1}{\sqrt{6}}\phi & K^{*0} \\ K^{*-} & \bar{K}^{*0} & -\sqrt{\frac{2}{3}}\phi \end{bmatrix} \quad (3.56)$$

We may construct SU_3 invariant interactions by taking traces of products of these matrices, which corresponds to contracting indices in a product of tensors of type (1, 1). Out of B and P, for example, we may construct Tr $(\bar{B}\,P\,B)$ and Tr $(\bar{B}\,B\,P)$, but there are no other distinct traces trilinear in these two matrices. The combinations Tr $(\bar{B}\,P\,B) \pm$ Tr $(\bar{B}\,B\,P)$ are know as $\frac{D}{F}$ - type couplings. However this quick method works only for the octets.

Isotopic Content of an Irreducible Representation

(This section is due to F. Pirani). Let us calculate which eigenstates of Y and I_3 are contained in any particular irreducible representation, and with what multiplicity. First we construct a basis for irreducible tensors of given type (p, q): recall that, because of the complete symmetry, a unique component of the irreducible tensor $T^{\beta_1 \dots \beta_q}_{\alpha_1 \dots \alpha_p}$ may be specified by giving the numbers of 1's, 2's and 3's among the indices, namely p_1, p_2, p_3 (lower indices, $\Sigma p_k = p$), and q_1, q_2, q_3 (upper indices, $\Sigma q_k = q$). This component may be denoted $T\begin{pmatrix} q_1 q_2 q_3 \\ p_1 p_2 p_3 \end{pmatrix}$. Also, the tensor is completely determined by the tracefree condition if the components which have $p_3 = 0$ or $q_3 = 0$, or both, (i.e., $p_3 q_3 = 0$) are specified; these components may be specified arbitrarily. Also every irreducible tensor has at least one such component. Therefore a basis for irreducible tensors of given types (p, q) may be constructed by enumerating those which have precisely one non-zero component with $p_3 q_3 = 0$. The other components of this tensor may be determined from the tracefree condition by removing pairs of 1's (or 2's) and inserting pairs of 2's (or 1's) and 3's. For example, the non-zero components of the tensor constructed from $T\begin{pmatrix} q_1 q_2 q_3 \\ p_1 p_2 0 \end{pmatrix}$ are all all of the form

$$T\begin{pmatrix} q_1 - r_1, & q_2 - r_2, & q_3 + r_1 + r_2 \\ p_1 - r_1 & p_2 - r_2, & r_1 + r_2 \end{pmatrix}. \tag{3.57}$$

The dimension of the basis is of course just the number of independent components of a tensor of type (p, q), namely $\frac{1}{2}(p + 1)(q + 1)(p + q + 2)$.

We know, from equation (3. 5), that each component of an irreducible tensor corresponds to a simultaneous eigenstate of operators corresponding to diagonal generators. The eigenvalue is

$$\sum_k n_k (p_k - q_k),$$

where n_k are the diagonal components of the generator. This eigenvalue depends only on differences of p's and q's. However, these differences are the same for all the non-vanishing components of a basis tensor, as can be seen from equation (3.57). This proves that every basis tensor corresponds to an eigenstate of the operator corresponding to any diagonal generator. The particular basis tensor may conveniently be identified by that component which has $p_3 q_3 = 0$. By counting the number of solutions of the equation

$$\lambda = \Sigma n_k (p_k - q_k) \tag{3.59}$$

by possible p's and q's, for given λ, we may determine the multiplicity of any eigenstate.

We now consider specifically the eigenstates of Y and I_3. From (3.38), we have

$$\text{for Y:} \quad n_1 = n_2 = \tfrac{1}{3}, \qquad n_3 = -\tfrac{2}{3}$$

$$\text{for } I_3: \quad n_1 = -n_2 = \tfrac{1}{2}, \quad n_3 = 0.$$

Therefore, for any eigenstate, (3.59) gives

$$Y = \tfrac{1}{3}(p_1 - q_1 + p_2 - q_2 - 2p_3 + 2q_3) = \tfrac{1}{3}(p - q) - (p_3 - q_3) \tag{3.60}$$

$$I_3 = \tfrac{1}{2}(p_1 - q_1 - p_2 + q_2) = p_1 - q_1 + \tfrac{1}{2}(p_3 - q_3) - \tfrac{1}{2}(p - q) \tag{3.61}$$

Equation (3.60) shows (as we have already seen) that only those tensors for which the numbers of upper and lower indices differ by a multiple of 3 will be states of integral hypercharge. Also it shows that interchange of upper and lower indices changes the sign of both Y and I_3. However, since the contravariant and covariant representations are not (except when p = q) equivalent, this "hypercharge reflection" is not an operation of SU_3.

Let us now count the distribution of eigenvalues among the basis tensors. It is convenient to work with integers, so we introduce

$$Y' = Y + \tfrac{2}{3}(p - q) = (p - q) - (p_3 - q_3) \tag{3.62}$$

Then (3.61) may be rewritten

$$p_1 = q_1 + (I_3 + \frac{1}{2} Y'). \qquad (3.63)$$

Note also that

$$\begin{aligned} 0 &\le p_3 \le p \\ 0 &\le q_3 \le q \end{aligned} \qquad (3.64)$$

and that for given p_3, q_3

$$\begin{aligned} 0 &\le p_1 \le p - p_3 \\ 0 &\le q_1 \le q - q_3 \end{aligned} \qquad (3.65)$$

Also if p_1, p_3, q_1, q_3 are given, satisfying $p_3 q_3 = 0$, then p_2 and q_2 are determined and a basis tensor is determined uniquely.

For $p_3 = 0$, (3.62) and (3.64) give

$$p_3 = 0 \Rightarrow p - q \le Y' \le p$$

and for $q_3 = 0$, (3.62) and (3.64) give

$$q_3 = 0 \Rightarrow -q \le Y' \le p - q \qquad (3.66)$$

Therefore

$$-q \le Y' \le p \qquad (3.57)$$

and if Y' is fixed in accord with (3.66), this fixes p_3, with $q_3 = 0$, or q_3, with $p_3 = 0$, according as $Y' \le p - q$ or $Y' \ge p - q$. For given Y', (3.62), (3.63), and (3.65) show that

$$\begin{aligned} Y' \le p - q &\Rightarrow -\left(q + \frac{1}{2} Y'\right) \le I_3 \le \left(q + \frac{1}{2} Y'\right) \\ Y' \ge p - q &\Rightarrow -\left(p - \frac{1}{2} Y'\right) \le I_3 \le \left(p - \frac{1}{2} Y'\right) \end{aligned} \qquad (3.68)$$

For any given Y' and I_3 consistent with (3.67) and (3.68), the number of independent basis tensors is just the number of solutions p_1, q_1 of equation (3.63) which satisfy also equation (3.65). It is convenient to treat the two cases $Y' \lesseqgtr p - q$ separately in counting these solutions.

Remark first that if a, b, c, d are integers, $a \le b$, $c \le d$, represented, say, by points on the real axis, then if the interval ab and the interval cd overlap, the length of overlap is min(b, d) - max(a, c) and the number of integer points of overlap is one more than this.

Case 1. $p_3 = 0$, $q_3 \ge 0$, $Y' \ge p - q$.

$$(3.65) \Rightarrow 0 \le p_1 \le p$$

$$0 \le q_1 \le q - q_3 = p - Y'$$

$$\Rightarrow (I_3 + \frac{1}{2} Y') \le q_1 + (I_3 + \frac{1}{2} Y') \le I_3 - \frac{1}{2} Y' + p$$

Hence multiplicity is

$$\mu_1 = \min(p, I_3 - \frac{1}{2} Y' + p) - \max(0, I_3 + \frac{1}{2} Y') + 1$$

$$= \min(0, I_3 - \frac{1}{2} Y') - \max(0, I_3 + \frac{1}{2} Y') + p + 1 \quad (3.69)$$

provided that this number is positive, that the value of Y' is possible (according to (3.66)) and that the value of I_3 is possible (according to (3.68)).

Case 2. $q_3 = 0$, $p_3 \ge 0$. $Y' \le p - q$.

$$(3.65) \Rightarrow 0 \le q_1 \le q$$

$$0 \le p_1 \le p - p_3 = q + Y'$$

$$\Rightarrow I_3 + \frac{1}{2} Y' \le q_1 + (I_3 + \frac{1}{2} Y') \le I_3 + \frac{1}{2} Y' + q.$$

Hence multiplicity is

$$\mu_2 = \min(q + I_3 + \frac{1}{2} Y', q + Y') - \max(0, I_3 + \frac{1}{2} Y') + 1$$

$$= \min(0, I_3 - \frac{1}{2} Y') - \max(0, I_3 + \frac{1}{2} Y') + q + Y' + 1, \quad (3.70)$$

under the same conditions as above.

We notice that the multiplicities given by (3.69) and (3.70) agree for $Y' = p - q$, as they should. Moreover, since $\min(0, a) = -\max(0, -a)$, each formula is unchanged by the substitution $I_3 \to -I_3$.

It is convenient to display the eigenvalues and multiplicities on an I_3 - Y graph, which is called a weight diagram (the name is taken over from group theory). This may easily be drawn with the help of the formals (3.66) - (3.70). We know already that it is symmetrical about the line $I_3 = 0$. The arrangement of occupied points, starting from the top row, is seen to be

Y'	I_3	No. of occupied points
p	$-\frac{1}{2}p, -\frac{1}{2}p+1, \ldots, \frac{1}{2}p$	$p+1$
$p-1$	$-\frac{1}{2}(p+1), -\frac{1}{2}(p+1)+1, \ldots, \frac{1}{2}(p+1)$	$p+2$
$\vdots$		
$p-q+1$	$-\frac{1}{2}(p+q-1), \ldots, \frac{1}{2}(p+q-1)$	$p+q$
$p-q$	$-\frac{1}{2}(p+q), \ldots, \frac{1}{2}(p+q)$	$p+q+1$
$p-q-1$	$-\frac{1}{2}(p+q-1), \ldots, \frac{1}{2}(p+q-1)$	$p+q$
$\vdots$		
$-(q-1)$	$-\frac{1}{2}(q+1), \ldots, \frac{1}{2}(q+1)$	$q+2$
$-q$	$-\frac{1}{2}q, \ldots, \frac{1}{2}q$	$q+1$

Thus the diagram widens a point at a time as Y' decreases from p to p - q, then narrows again as Y' continues to decrease from p - q to -q.

Equations (3.69) and (3.70) show that loci of constant multiplicity are determined by the signs of $Y' - (p-q)$, $I_3 + 1/2\,Y'$, and $I_3 - 1/2\,Y'$. The following table gives the possible cases $\left(\text{but} \begin{Bmatrix} - + - \\ + - + \end{Bmatrix} \text{ occurs only for } p - q \begin{Bmatrix} >0 \\ <0 \end{Bmatrix}\right)$

Sign of $Y' - (p - q)$	Sign of $I_3 + \frac{1}{2} Y'$	Sign of $I_3 - \frac{1}{2} Y'$	Loci of constant multiplicity
+	+	+	$I_3 + \frac{1}{2} Y' = \text{const.}$
+	+	-	$Y' = \text{const.}$
+	-	+	Whole region
+	-	-	$I_3 - \frac{1}{2} Y' = \text{const.}$
-	+	+	$I_3 - \frac{1}{2} Y' = \text{const.}$
-	+	-	Whole region
-	-	+	$Y' = \text{const.}$
-	-	-	$I_3 + \frac{1}{2} Y' = \text{const.}$

With all this information it is easy to sketch the general appearance of the weight diagram: If the lines $I_3 = \pm 1/2\, Y'$, $Y' = p - q$ are drawn, they enclose (unless concurrent) a triangle of constant multiplicity, surrounded by hexagons of constant multiplicity:

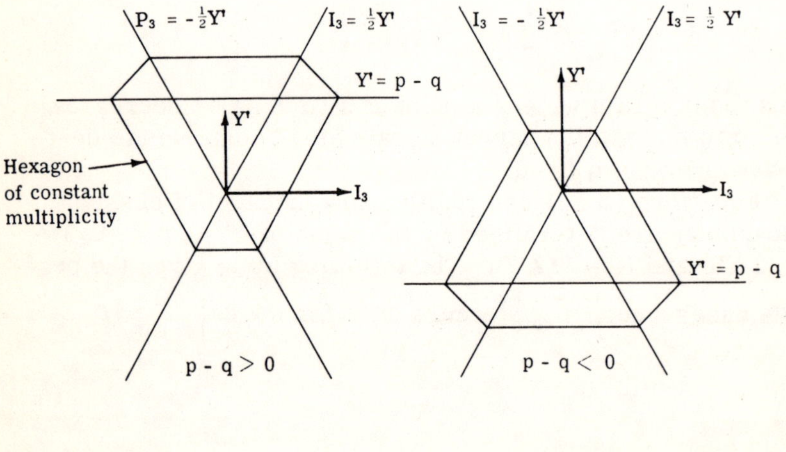

Here, for example, are the weight diagrams for (4, 1) and (2, 2):

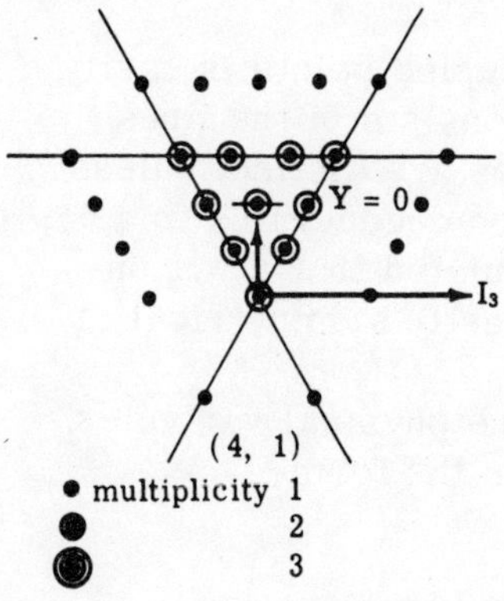

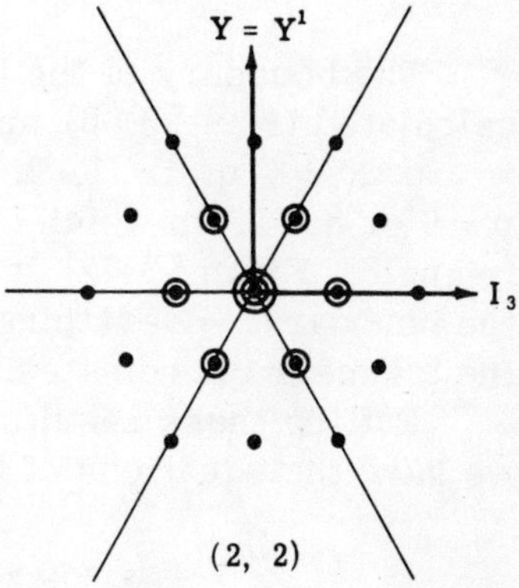

How can we bring into evidence the triangular symmetry which is apparent in these diagrams? One simple way is as follows: First change the scale of I_3 to make the triangle equilateral. A suitable substitution is $X = 2/\sqrt{3}\, I_3$. Since I_3 is an integer or half integer at each occupied point $X = m/\sqrt{3}$, where m is an integer. The lines $I_3 = \pm 1/2\, Y'$ become $X = \pm 1/\sqrt{3}\, Y'$. These lines and the line $Y' = p - q$ form an equilateral triangle. Take the sides of this triangle as bases for trilinear coordinates x_1, x_2, x_3 (which represent perpendicular distances from the sides, distance measured into the half-plane containing the center of the triangle being counted positive). Those coordinates are

$$\left.\begin{aligned} x_1 &= \frac{1}{2}(Y' + \sqrt{3}\,X) \\ x_2 &= \frac{1}{2}(Y' - \sqrt{3}\,X) \\ x_3 &= (p - q) - Y', \text{ with} \\ x_1 + x_2 + x_3 &= p - q \end{aligned}\right\} \qquad (3.71)$$

It is easy to see that each of these coordinates takes only integer values at the occupied points of the weight diagram.

Moreover, in terms of them, the two multiplicity formulae (3.69) and (3.70) may be combined into a single formula

$$\mu = p + 1 - \max(0, x_1) - \max(0, x_2) - \max(0, x_3) \quad (3.72)$$

The boundary of the hexagon of occupied points is easily calculated from (3.66) and (3.68): it consists of the lines $x_i = p$, $x_i = -q$, $i = 1, 2, 3$. These lines are distinct unless $p = 0$ or $q = 0$, in which case the hexagon degenerates to a triangle. From (3.72) it is easily calculated that $\mu = 1$ on the boundary. Everything is now manifestly symmetrical in the trilinear coordinates.

Putting these results in terms of the physical variables, we have the equations of the boundary in the form

$$\left.\begin{aligned}
\frac{1}{2}Y + I_3 &= p - \frac{1}{3}(p-q)\\
\frac{1}{2}Y + I_3 &= -q - \frac{1}{3}(p-q)\\
\frac{1}{2}Y - I_3 &= p - \frac{1}{3}(p-q)\\
-\frac{1}{2}Y - I_3 &= -q - \frac{1}{3}(p-q)\\
Y &= -p + \frac{1}{3}(p-q)\\
Y &= q + \frac{1}{3}(p-q)
\end{aligned}\right\} \quad (3.73)$$

and the multiplicity formula

$$\begin{aligned}
\mu = p+1 &- \max\left(0, \frac{1}{2}Y + I_3 + \frac{1}{3}(p-q)\right)\\
&- \max\left(0, +\frac{1}{2}Y - I_3 + \frac{1}{3}(p-q)\right)\\
&- \max\left(0, \frac{1}{3}(p-q) - Y\right)
\end{aligned} \quad (3.74)$$

Reduction of the Direct Product Again

Next I should like to give another method for the reduction of the direct product, without the use of Young diagrams. Suppose we are given two SU_3 - tensors $T_{\alpha_1 \dots \alpha_p}^{\beta_1 \dots \beta_q}$ and $Q_{\alpha'_1 \dots \alpha'_p}^{\beta'_1 \dots \beta'_q}$. How can the product of these two be expressed as the sum of irreducible tensors? We turn this question around by asking: what independent invariants can be formed from the product of these two with an arbitrary tensor $S_{\beta_1 \dots}^{\alpha_1 \dots}$?

Consider for example the product

$$T_\alpha^\beta \quad Q_\delta^\gamma.$$

First of all, we can form an invariant by multiplication with arbitrary (traceless, symmetric) irreducible tensor

$$\underset{1}{S}{}_{\beta\gamma}^{\alpha\delta}.$$

Next, we can form an invariant by lowering β and γ and then multiplying by an arbitrary tensor:

$$T_\alpha^\beta \; Q_\delta^\gamma \; \epsilon_{\beta\gamma\mu} \; \underset{2}{S}{}^{\alpha\delta\mu}.$$

Continuing in this way, we can form the following invariants:

$$T_\alpha^\beta \, Q_\delta^\gamma \times \underset{1}{S}{}_{\beta\gamma}^{\alpha\delta} \qquad (2,\ 2) = 27$$

$$\times\, \epsilon_{\beta\gamma\mu} \, \underset{2}{S}{}^{\alpha\delta\mu} \qquad (0,\ 3) = \overline{10}$$

$$\times\, \epsilon^{\alpha\delta\mu} \, \underset{3}{S}{}_{\beta\gamma\mu} \qquad (3,\ 0) = 10$$

$$\times\, \delta_\gamma^\alpha \, \underset{4}{S}{}_\beta^\delta \qquad (1,\ 1) = 8$$

$$\times\, \delta_\beta^\delta \, \underset{4}{S}{}_\gamma^\alpha \qquad (1,\ 1) = 8$$

$$\times\, \delta_\gamma^\alpha \, \delta_\beta^\delta \, S_5 \qquad (0,\ 0) = 1.$$

Note that we get nothing new if we both raise lower indices and lower upper indices with ϵ's because a product of two ϵ's is just a linear combination of Kronecker deltas which gives nothing new:

$$\epsilon_{\alpha_1\alpha_2\alpha_3}\cdot\epsilon^{\beta_1\beta_2\beta_3} = \begin{vmatrix} \delta^{\beta_1}_{\alpha_1} & \delta^{\beta_2}_{\alpha_1} & \delta^{\beta_3}_{\alpha_1} \\ \delta^{\beta_1}_{\alpha_2} & \delta^{\beta_2}_{\alpha_2} & \delta^{\beta_3}_{\alpha_2} \\ \delta^{\beta_1}_{\alpha_3} & \delta^{\beta_2}_{\alpha_3} & \delta^{\beta_3}_{\alpha_3} \end{vmatrix} = \delta^{\beta_1}_{\alpha_1}\,\delta^{\beta_2}_{\alpha_2}\,\delta^{\beta_3}_{\alpha_3} + \text{five similar terms.}$$

In the two (1, 1) invariants it is convenient (but not compulsory) to remove the trace from the product before forming the invariant, for example

$$T^{\beta}_{\alpha}\, Q^{\gamma}_{\delta} - \frac{1}{3}\,\delta^{\beta}_{\delta}\, T^{\mu}_{\nu}\, Q^{\nu}_{\mu} \times \delta^{\alpha}_{\gamma}\; S^{\delta}_{4\,\beta}$$

instead of the first (1, 1) case.

Again we see the result

$$8 \otimes 8 = 27 \oplus 10 \oplus \overline{10} \oplus 8 \oplus 8 \oplus 1.$$

In the general case, we carry out the same procedure in a systematic way. We start with

$$T^{\beta_1\cdots\beta_q}_{\alpha_1\cdots\alpha_p} \qquad Q^{\beta'_1\cdots\beta'_q}_{\alpha'_1\cdots\alpha'_p}$$

and

1) multiply by $S^{\alpha_1\cdots\alpha_p\alpha'_1\cdots\alpha'_p}_{\beta_1\cdots\beta_q\beta'_1\cdots\beta'_q}$

2) ϵ in all possible ways, then multiply by an S:

$$\epsilon_{\alpha_1\alpha'_1\mu_1},\ \epsilon_{\alpha_2\alpha'_2\mu_2},\ \ldots \text{ up to min (p, p') times}$$

$$\epsilon^{\beta_1\beta'_1\rho_1},\ \epsilon^{\beta_2\beta'_2\rho_2}\ \ldots \text{ up to min (q, q') times}$$

After this ϵ'ing, no more tracing is possible (traces vanish).

3) for each possible way of ϵ'ing,

Trace

$$\alpha_1 \text{ and } \beta'_1,\ \alpha_2 \text{ and } \beta'_2 \ \ldots \text{ up to min } (p,\ q') \text{ times}$$

and

$$\alpha'_1 \text{ and } \beta_1,\ \alpha'_2 \text{ and } \beta_2 \ \ldots \text{ up to min } (p',\ q) \text{ times.}$$

Exercises: $27 \times 10 = ?$
$27 \times 35 = ?$

The advantage of this method is that it gives the invariants explicitly, and so if the theory is right you have the coupling coefficients and you can immediately calculate the matrix elements between given states, and find decay widths and so on. For example if the width of $N^* \to N + \pi$ is given, then, putting the masses in by hand, you can predict

$$Y_1^*(1385) \to \begin{matrix} \Sigma + \pi \\ \Lambda + \pi \end{matrix} \text{ and } \Xi^*(1530) \to \Xi + \pi.$$

The $\Lambda + \pi$ and $\Xi + \pi$ widths come out correctly from this simple model, but the $\Sigma + \pi$ width is a bit too small.

I spin, U spin and V spin

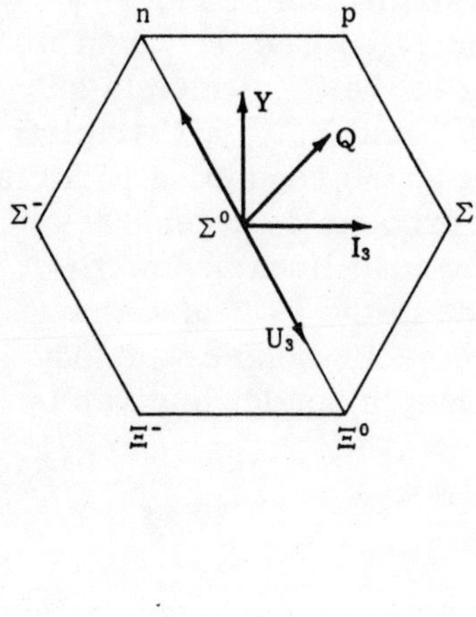

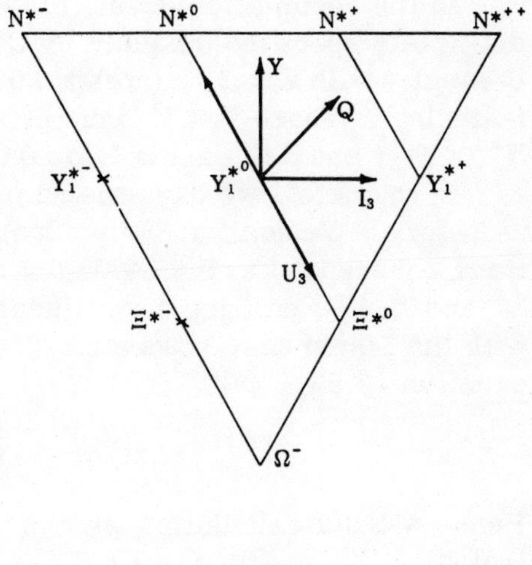

Now I wish to discuss the ingenious observation, first made by Meshkov, Levinson and Lipkin[24] that the SU_3 group itself does not distinguish any particular direction in the space on which it acts. We have picked out the 3-axis for special treatment by our choice of generators (3.38) and by the physical interpretation, which makes $T_3^3 = -Y$, and then the 1 and 2 directions correspond to I-spin. From this point of view one looks at the weight diagrams in the orientation in which they are drawn above, with the horizontal and vertical axes picked out. The states of an isotopic spin multiplet lie on the same horizontal line.

But because of the symmetry of the diagrams, one could just as well pick out another axis for preferred treatment, and there is another one which is interesting from the physical point of view, and that is the 1-axis, because $T_1^1 = Q$ (equation (3.77)). In effect we rotate the diagram (or the axes) through 120°. Then the 2 and 3 directions correspond to a new "spin" variable, which Lipkin et al, call U-spin. Similarly we could pick out the 2-axis for special treatment, and then the 1 and 3 directions correspond to the so-called V-spin, but this seems to be physically less interesting.

Let us see what we get with this new point of view: U-spin is quite closely analogous to I-spin.

In the decuplet diagram, for example, we see four multiplets specified uniquely by Q and U_3. The N^{*++} is a U-singlet with Q = 2 (corresponding to the Ω^-, which is an I-singlet). There is a U-doublet: N^{*+} and Y_1^{*+}, a U-triplet: N^{*0}, Y_1^{*0} and Ξ^{*0}, and a U-quartet of the remaining particles.

In the octet, we have the same problem of degeneracy as before. We don't know in advance what linear combinations correspond to the U-singlet and to the U-triplet among Σ^0 and Λ. We can get these linear combinations by analogy with the I-spin case however. The I-spin singlet is given by equation (3.53):

$$M_3^3 = -\sqrt{\frac{2}{3}}\,\Lambda$$

Hence without calculation we can infer (with a sign convention) that

$$M_1^1 = +\sqrt{\frac{2}{3}}\,(\text{U singlet}).$$

But we have from (3.54)

$$M_1^1 = \frac{1}{\sqrt{2}}\Sigma^0 + \frac{1}{\sqrt{6}}\Lambda^0$$

Therefore

$$U \text{ singlet} = \frac{\sqrt{3}}{2}\Sigma^0 + \frac{1}{2}\Lambda^0.$$

The triplet has to be orthogonal to the singlet, for example

$$U \text{ triplet} = \frac{1}{2}\Sigma^0 - \frac{\sqrt{3}}{2}\Lambda^0.$$

The selection rules for processes involving these mixed states will not be very interesting, because they will involve complex amplitudes, which means that we will only get triangle inequalities; if we can find processes involving only pure states we will get stronger conditions. The U-spin formalism is very nice for deriving selection rules for strong interactions, as Lipkin et al have shown.

If you are looking at actual scattering interactions, it is difficult to find a criterion for comparision of experiments, because normally you have to compare experiments in different channels. Sometimes you can compare experiments in one channel, with different decay modes. You have to stay away from resonances and thresholds. For example, if you want to compare $\bar{K}$ - N scattering with π - N scattering you can get into serious trouble if you are near a resonance. It is much the simplest if you can stay in one channel. U-spin gives you a convenient way of making comparisons, especially if you start with a U-spin singlet. There is a particular U-spin singlet in nature which can be used to initiate experiments, namely the photon. The simplest candidates for such experiments are the neutron (which is in a U-spin triplet) and the proton (which is in a U-spin doublet). Consider for example

$$\begin{array}{lcll} U = \frac{1}{2} & & U = 1 & U = \frac{1}{2} \\ \hline \gamma + p & \rightarrow & N^{*0} & + \pi^+ \\ & \rightarrow & Y_1^{*0} & = K^+ \end{array}$$

SU_3-invariance requires that we conserve U-spin. The combination of amplitudes which gives $U = \frac{3}{2}$ is excluded. The cross-sections for these two processes are proportional to expressions of the form

$$|(1,\ 1;\ \tfrac{1}{2},\ -\tfrac{1}{2}|\ \tfrac{1}{2},\ \tfrac{1}{2})|^2\ \ |M|^2$$

$$|(1,\ 0;\ \tfrac{1}{2},\ \tfrac{1}{2}|\ \tfrac{1}{2},\ \tfrac{1}{2})|^2\ \ |M|^2;$$

the Clebsch-Gordan coefficients can be formed from tables and the other factor has the same value in the two cases.

Exercise: Suppose there is a $\overline{10}$ baryon:

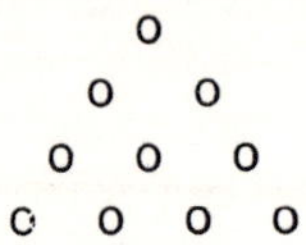

The I-spin doublet has $Y = 1$, $I = \frac{1}{2}$. Show that $\gamma + p \nrightarrow \overline{10}$ is forbidden by U-spin conservation.

A possible example of this is the 2190 Mev π p resonance with $I = \frac{1}{2}$. This might be a recurrence of the nucleon with spin $\frac{9}{2}$, but the spin is not yet reliably known. Osborne and Bar Yam have proposed an experiment to check whether this is the $\overline{10}$ or not. It does appear in πp scattering, and should appear also in γn scattering but not in γp scattering.

Trouble with Experimental Verification of SU_3.

Let me suggest to you a difficulty (which arises with any strongly broken symmetry) in checking SU_3 at high energies. The above exercise gives one example (but there are others). Consider this interaction (see illustration on top of following page).

There is a definite ratio in the cross-section. We know that for low mass we don't get that ratio. But in the high mass limit we don't get it, either: no-one would argue that the vector octet particles, for example, ρ, ϕ and K* are in any basic sense elementary particles. Presumably they are to be understood in terms of dynamical interactions of 2π's or πK,

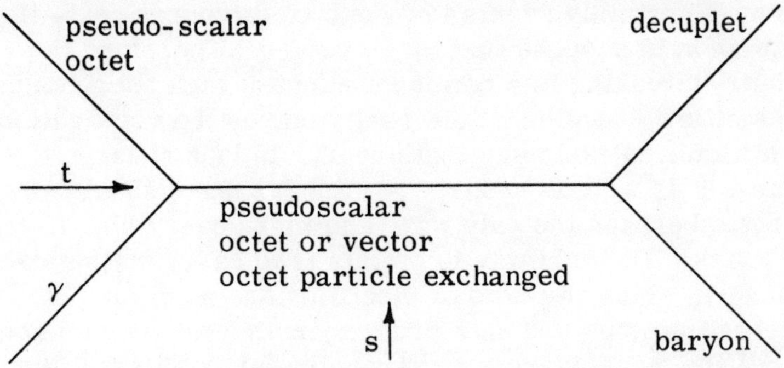

or 2K's or something like that. Therefore the appropriate way to calculate the exchange is not with elementary particle exchange but with Regge exchange. If the reaction yields $N^* + \pi$, we have ρ exchanged; if it yields $Y_1^* + K$, we have K^* exchanged (by SU_3 symmetry). The Regge mechanism predicts that at high energies the energy dependence of the cross-section, up to factors which tend to 1, is given by $s^{\alpha_\rho(t)}$, and by $s^{\alpha_{K^*}(t)}$ respectively. Therefore the forward or near forward ratio is

$$\frac{d\sigma}{d\Omega_1} \bigg/ \frac{d\sigma}{d\Omega_2} \sim s^{\alpha_\rho - \alpha_{K^*}} \sim s^{1/4}, \text{ roughly,}$$

as we can predict from this graph:

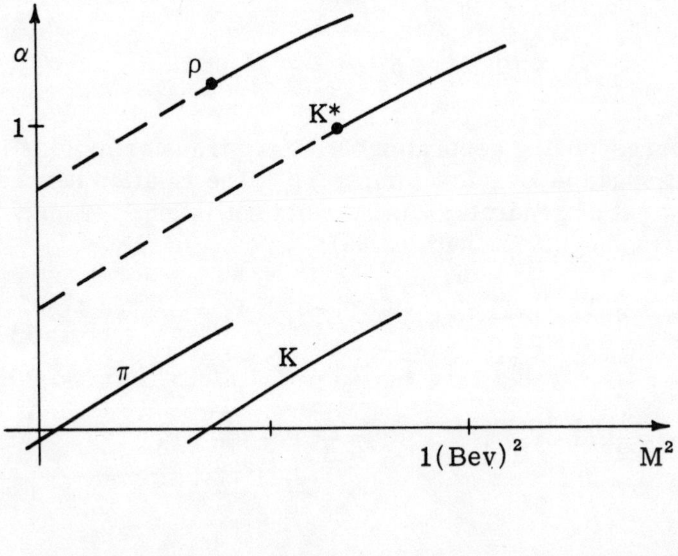

Anyhow, s certainly seems to have a positive exponent—there is no reason to suppose that $\alpha_\rho - \alpha_{K^*} \to 0$. Therefore the symmetry-breaking is a dominant effect at high energies as well as at low energies. This really makes the verification very difficult. People say that one should look at large t and large u as well as at large s. But the mass should also be large. Perhaps the only way is to make everything have large mass. The only way to do this is to carry out an experiment in which two sets of electrons and positions or two sets of protons and anti-protons annihilate, or come out. Then all the masses are large (negative or positive). But the experimentalists would find this a little hard to do, and the question has to be left somewhat up in the air.

Mass Formula

This is the real basis for believing that SU_3 works. The assumption is that the mass splitting transforms like $T_8 = Y$.

In order to develop the mass formula, it is convenient to introduce a second set of generators in place of $T_1 \ldots T_8$ (defined by (3.38)). We take generators of the SU_3 transformations which have just one non-zero component, except that if that component lies on the diagonal we subtract a multiple of the unit tensor in order to make the generator trace free. Thus we define T^ν_μ by

$$(T^\nu_\mu)^\beta_\alpha = \delta^\nu_\alpha \, \delta^\beta_\mu - \frac{1}{3}\,\delta^\nu_\mu \, \delta^\beta_\alpha. \qquad (3.75)$$

The corresponding generator of unitary transformations in Hilbert space is denoted simply T^ν_μ. The relation between the new set of generators and the old set is found from a comparison of (3.75) with (3.38):

$$\left.\begin{array}{lll} T_1 = \frac{1}{2}(T^1_1 + T^2_2) & T_4 = \frac{1}{2}(T^1_3 + T^3_1) & T_6 = \frac{1}{2}(T^2_3 + T^3_2) \\ T_2 = \frac{1}{2}i(T^1_2 - T^2_1) & T_5 = \frac{1}{2}i(T^1_3 - T^3_1) & T_7 = \frac{1}{2}i(T^2_3 - T^3_2) \\ T_3 = \frac{1}{2}(T^1_1 - T^2_2) & T_8 = \frac{1}{3}(T^1_1 + T^2_2 - 2T^3_3) & \end{array}\right\} \qquad (3.76)$$

Although it would appear on first sight that nine generators are defined by (3.75), in fact only eight of these are independent, the diagonal generators satisfy the relation

$$T_{\mu}^{\mu} = 0. \tag{3.77}$$

From the definition (3.75) we find the commutation relations between the new generators:

$$([T_{\mu}^{\nu}, T_{\rho}^{\sigma}])_{\beta}^{\alpha} = (T_{\mu}^{\nu})_{\gamma}^{\alpha} (T_{\rho}^{\sigma})_{\beta}^{\gamma} - (T_{\rho}^{\sigma})_{\gamma}^{\alpha} (T_{\mu}^{\nu})_{\beta}^{\gamma} \qquad (3.78)$$

$$= (\delta_{\rho}^{\nu} T_{\mu}^{\sigma} - \delta_{\mu}^{\sigma} T_{\rho}^{\nu})_{\beta}^{\alpha},$$

or, for the generators in Hilbert space,

$$[T_{\mu}^{\nu}, T_{\rho}^{\sigma}] = \delta_{\rho}^{\nu} T_{\mu}^{\sigma} - \delta_{\mu}^{\sigma} T_{\rho}^{\nu}. \qquad (3.78)$$

On the other hand, if D_{α}^{β} is any irreducible tensor operator of type (1, 1), we have from (3.43)

$$[T_{\mu}^{\nu}, D_{\alpha}^{\beta}] = (T_{\mu}^{\nu})_{\alpha}^{\gamma} D_{\gamma}^{\beta} - (T_{\mu}^{\nu})_{\delta}^{\beta} D_{\alpha}^{\delta}$$

$$= \delta_{\alpha}^{\nu} D_{\mu}^{\beta} - \delta_{\mu}^{\beta} D_{\alpha}^{\nu}. \tag{3.79}$$

From an operator wave function $\psi_{\alpha_1 \ldots \alpha_p}^{\beta_1 \ldots \beta_q}(x)$ of arbitrary type (p, q), we can construct two different tensor operators of type (1, 1) with the aim of calculating a mass formula, namely

$$\int dx \, \{\bar{\psi}_{\delta_1 \ldots \delta_q}^{\gamma_1 \ldots \gamma_{p-1}\alpha} \, \psi_{\gamma_1 \ldots \gamma_{p-1}\beta}^{\delta_1 \ldots \delta_q} - \delta_{\beta}^{\alpha} \, (\text{trace})\}$$

and (3.80)

$$\int dx \, \{\bar{\psi}_{\delta_1 \ldots \delta_{q-1}\beta}^{\gamma_1 \ldots \gamma_p} \, \psi_{\gamma_1 \ldots \gamma_p}^{\delta_1 \ldots \delta_{q-1}\alpha} - \delta_{\beta}^{\alpha} \, (\text{trace})\},$$

but it is more convenient to have a formula from which the masses can be read off directly. Whenever both the integrals (3.80) can be written down (i.e., except when p = 0 or q = 0), there should be two independent mass operators. We can take $T_3^3 = Y$ for one of them. The other one must transform in the same way but be independent of T^3_3. The $^3_3$ component of $T^\mu_\sigma T^\sigma_\nu - T^\sigma_\nu T^\mu_\sigma$ is evidently unsatisfactory; because (3.78) tells us that it is proportional to T^μ_ν, and therefore it gives nothing new. Consider on the other hand the operators D^μ_ν, defined by

Casimir operator

$$-2D^\mu_\nu = T^\mu_\sigma T^\sigma_\nu + T^\sigma_\nu T^\mu_\sigma - \frac{2}{3}\delta^\mu_\nu T^\rho_\sigma T^\sigma_\rho . \tag{3.81}$$

Conventional — Removes trace

We shall put this operator into a form which shows explicitly that it is independent of T^μ_ν. First we calculate the Casimir operator

$$C = T^\rho_\sigma T^\sigma_\rho = (T^3_3)^2 + T^3_i T^i_3 + T^i_3 T^3_i + T^i_j T^j_i \quad (i,\ j = 1, 2)$$

which may be rearranged into

$$C = -(T^3_3)^2 + (T^3_\sigma T^\sigma_3 + T^\sigma_3 T^3_\sigma - \frac{2}{3} C) + \frac{2}{3} C + T^i_j T^j_i$$

and with the help of (3.76) we have

$$\left.\begin{aligned} T^1_1 &= I_3 + \frac{1}{2} Y, \\ T^2_2 &= -I_3 + \frac{1}{2} Y, \\ T^3_3 &= -Y, \end{aligned}\right\} \tag{3.82}$$

so that

$$\begin{aligned} \frac{1}{3} C &= -\frac{1}{2}Y^2 - 2D^3_3 + 2I_3{}^2 + T^2_1 T^1_2 + T^1_2 T^2_1 \\ &= -\frac{1}{2} Y^2 - 2D^3_3 + 2\vec{I}^{\,2}, \end{aligned} \tag{3.83}$$

(since the left hand side is invariant, and the first two terms on the right hand side of (3.83) are invariant under isotopic spin rotations, so must the last three terms be, which shortens the calculation).

Hence

$$D_3^3 = \vec{I}^2 - \frac{1}{4} Y^2 - \frac{1}{6} C. \qquad (3.82)$$

Exercise: Verify that $C = 2(p+q) + 2/3\,(p^2 + pq + q^2)$.

Hint: Calculate

$$[T_\sigma^\rho, [T_\rho^\sigma, T_{\alpha_1 \ldots \alpha_p}^{\beta_1 \ldots \beta_q}]]$$

with the help of (3.43).

From (3.81) and (3.82) we may now write down the Gell-Mann-Okubo mass formula

$$\delta M = m_1\, Y + m_2\, D_3^3. \qquad (3.83)$$

Except when $p = 0$ or $q = 0$, Y and D_3^3 are linearly independent.

Discussion of the Mass Formula.

The mass formula works much better than it has any conceivable right to work. For the pion octet, it works to within 10%. You do even better if you use the squared masses in the mass formula, but the justification for doing that is not terribly clear. The formula works almost perfectly for the baryon octet, and to the accuracy of isotopic spin breaking of assumed electromagnetic shift for the decuplet. In the case of the decuplet there seems to be an accident happening, because we understand enough of the dynamics to be confident in considering for example the N^* to be a dynamically resonant state of π and N, and if that is the case than the Wigner effect, which I believe is what Yang and Oakes were thinking of, comes in. That is to say, if there are closed channels and open channels and states with large widths, then the location of the resonance depends very definitely on which channels are closed and which are open. One thinks always in terms of a formula like

$$\frac{1}{\omega_R - \omega - i \sum_c \gamma_c q_c^3} \qquad \left(\sum_c \text{ is } \sum_{\text{channels}}\right)$$

(q^3 since this is a P - state). We can list the open and closed channels, for example

	Open	Closed
1238	πn	$K\Sigma$
1385	$\Lambda\pi$, $\Sigma\pi$	$\Sigma\eta$, $N\bar{K}$

There may be other channels. The point is that some channels are closed and some are open, and whether they are closed or open is an accidental consequence of what the mass formula chooses to do. In a closed channel q becomes imaginary and therefore contributes to ω_R. This is a standard effect which is known to exist and which sometimes conceals the conservation of isotopic spin in nuclear physics. If there is isotopic spin conservation then there are levels which are just bound or just unbound, but here they get shifted by a substantial accidental amount. The width is not a symmetry-breaking effect, but the shift can move one up to become real, or imaginary. What is the order of magnitude? The 3-3 width is 100 MeV or so. According to SU_3 half of the width is a K Σ, which probably isn't there, so the total width is equivalent to 200 MeV. Thus, random contributions of 100 Mev or so are floating around. Yang and Oakes[18] had a much more detailed argument which turned out to be incorrect, but I think that they were trying to say just this in a precise way. The first counter example was a detailed dynamical calculation by Martin and Wali , who in fact found equal spacing, so that the effect had not shown up. This was actually before Yang and Oakes; afterwards a number of other people repeated the calculation in much simpler language. You find, remarkably, that the troublesome terms cancel out. This is not included in any of this formalism, and not to be understood by the various explanations of octet splitting, such as that of Dashen and Frautschi[12], which are perturbative. This is not a perturbative effect.

ϕ - ω Mixing.

The worst prediction of the mass formula is in the case of the vector octet. Here we have $\rho(750)$ and $K^*(890)$. But where is the ϕ (the isotopic scalar)? The V_8 particle should appear at about 930 MeV, but there there is no particle with the appropriate properties. Two particles which would fit very well are the $\omega(780)$ and the $\phi(1020)$. You can say: suppose that there had been a member of an octet, and a SU_3 scalar, which were even closer, and which were mixed by the interaction, which has off diagonal matrix elements and could connect an octet to a singlet. If you assume that ω and ϕ are mixtures of V_8 and a singlet V_1, then you know the masses and you get a unique mixing angle. The centre of mass is about 900, so the V_1, is about 870. As long as the V_8 mass is between the ϕ and ω masses you can fit the V_1, because the mixing always pushes the levels apart. I do not think that the mixing is very significant. To study mixing, we might consider the (very wrong) integral charge quark model: here the mixing takes place, but is not necessarily very relevant. We have p, n and λ to construct particles, and $\bar{p}$, $\bar{n}$, $\bar{\lambda}$ to construct antiparticles, and we put them together to make bosons. The ρ is an I = 1 state, the ω an I = 0 state, and together they exhaust the four ${}^3S$ states of nucleon-antinucleon. According to this model we have

$$\omega = \frac{1}{\sqrt{2}} (\bar{p}\,p + \bar{n}\,n)$$

$$\rho^+ = p\,\bar{n}$$

$$\rho^0 = \frac{1}{\sqrt{2}} (\bar{p}\,p - \bar{n}\,n)$$

$$\rho^- = n\,\bar{p}$$

and

$$\phi = \bar{\lambda}\lambda,\ K^{*+} = p\bar{\lambda},\ K^{*0} = n\bar{\lambda},\ K^{*-} = \bar{p}\lambda,\ \bar{K}^{*0} = \bar{n}\lambda.$$

(here we use complex conjugate, not charge conjugate wavefunctions). In this model we can calculate masses by simple addition. If we take m_ρ and m_ω to be equal, then $m_\lambda - m_p = m_{K^*} - m_\rho \sim 140$ MeV, and similarly $m_\phi - m_{K^*} \sim 130$ MeV, so that this model must therefore (accidentally) describe the mixing properties of the system. Now we write the SU_3 states: We have for the unmixed states:

$$V_1 = \frac{1}{\sqrt{3}}(\bar{p}\,p + \bar{n}\,n + \bar{\lambda}\,\lambda)$$

and

$$V_8 = \frac{1}{\sqrt{6}}(\bar{p}\,p + \bar{n}\,n - 2\bar{\lambda}\,\lambda)$$

So we have a mixing model, but it is in no way clear that the unmixed V states are more fundamental than the mixed states.

Final Exercise on SU_3:

Calculate the magnetic moments of all particles in terms of μ_p and μ_n, using U spin and the mass formula.

REFERENCES

1. This topic is a revised and expanded version of the first part of the lecturer's Lectures on Elementary Particles and Scattering Theory, Tata Institute of Fundamental Research, Bombay, 1964.
2. T.D. Lee and C.N. Yang, Phys. Rev., **98**, 1501 (1955). See also Y. Fujii, Prog. Theor. Phys., **21**, 232 (1959).
3. J.J. Sakurai, Ann. Phys., **11**, 1 (1960).
4. Christenson, Cronin, Fitch, and Turlay, Phys. Rev. Lett., **13**, 138 (1964).
5. The first experiment is that by Wu et al., Phys. Rev., **105**, 1413 (1957).
6. Beam and the Experiments on (Qp - Qe)/Q: J.C. Zorn et al., Proc. 1960 Rochester Conf.
7. R.H. Dicke, R. Krotkov, P.G. Roll, Annals of Physics, **26**, 442 (1964).
8. J. Schwinger, 1964 Brandeis Lectures (Prentice-Hall).
9. G. Feinberg and M. Goldhaber, Proc. N.A.S., **45**, 1301 (1959).
10. J.J. Sakurai, Phys. Rev. Lett., **10**, 446-448 (1963).
11. A.H. Rosenfeld et al., UCRL-8030, Part 1 (March 1965).
12. R. Dashen and S. Frautschi, Phys. Rev., **135**, 1196 (1964).

13. R.H. Hildebrand, Phys. Rev., **84**, 581 (1951).
14. L. Michel, Nuovo Cimento, **10**, 319 (1953).
15. N. Cabibbo, Phys. Rev. Lett., **14**, 965-968 (1965).
16. J. Prentki and M. Veltman, Phys. Rev. Lett., **15**, 88 (1965).
17. J. Bernstein, G. Feinberg, and T.D. Lee, Phys. Rev., **139**, 1650 (1965).
18. C.N. Yang and R. Oakes, Phys. Rev. Lett., **11**, 174 (1963).
19. The standard reference is R.E. Behrends, J. Dreitlein, C. Fronsdal, and W. Lee, Rev. Mod. Phys., **34**, 1 (1962). Much of the material presented there comes from G. Racah, Group Theory and Spectroscopy, Princeton Lecture Notes soon to be published by Ergebinsse Der Naturwissenschaften. Other standard works are H. Weyl, The Classical Groups, Princeton, 1946, and D.E. Littlewood, The Theory of Group Characters. The approach of M.E. Meyer, Brandeis Summer Institute, 1963 Lecture Notes, has some kinship to the present work. Recent work on representations and useful references are given by G.E. Baird and L. C. Biedenham, J. Math. Phys., **4**, 436, 1449 (1963), **5**, 1723 1730 (1964). A very useful new book is H.J. Lipkin, Lie Groups for Pedestrians, North-Holland, 1965.
20. G.B. Gurentch, Foundations of the Theory of Algebraic Invariants, Nordhoff, 1964.
21. See H. Weyl, The Classical Groups.
22. Behrends et al.
23. D.E. Littlewood, The Theory of Group Characters.
24. S. Meshkov, C.A. Levinson, and H.J. Lipkin, Phys. Rev. Lett., **10**, 361-364 (1963).